Awareness of Information Sources

*Theodore B. Selover, Jr., and
Max Klein, editors*

C.W. Bale
Peter Barun
David F. Bergman
W.T. Cavanaugh
Joseph R. Downey
Robert D. Freeman
Bruce E. Gammon
N.A. Gokcen
James A. Graham
Loren G. Hepler

William H. Kirchoff
Peter E. Liley
Neil A. Olien
A.D. Pelton
Sidney L. Phillips
Harold J. Raveche
T.B. Selover, Jr.
John Shaw
Carl Sutton
W.T. Thompson

Jack H. Westbrook

AIChE Symposium Series

1984

Number 237

Volume 80

Published by
American Institute of Chemical Engineers

345 East 47 Street

New York, New York 10017

Copyright 1984

American Institute of Chemical Engineers
345 East 47 Street, New York, N.Y. 10017

AIChE shall not be responsible for statements or opinions advanced in papers or printed in its publications.

Library of Congress Cataloging in Publication Data
Main entry under title:

Awareness of information sources.

(AIChE symposium series ; no. 237, v. 80)
"Papers selected from the three sessions at the Denver (August 1983) AIChE meeting entitled 'Awareness of information sources'"—Foreword.
 1. Chemical engineering—Information services—Addresses, essays, lectures. I. Selover, Theodore B., 1931-   . II. Klein, Max, 1925-   . III. Series: AIChE symposium series ; no. 237.
TP149.A98 1984   660.2'07   84-18598
ISBN 0-8169-0326-3

Authorization to photocopy items for internal or personal use, or the internal or personal use of specific clients, is granted by AIChE for libraries and other users registered with the Copyright Clearance Center (CCC) Transactional Reporting Service, provided that the $2.00 fee per copy is paid directly to CCC, 21 Congress St., Salem, MA 01970. This consent does not extend to copying for general distribution, for advertising or promotional purposes, for inclusion in a publication, or for resale.

Articles published before 1978 are subject to the same copyright conditions and the fee is $2.00 for each article. AIChE Symposium Series fee code: 0065-8812/84 $2.00

Printed in the United States of America by
Twin Production & Design

FOREWORD

This volume presents papers selected from the three sessions at the Denver (August 1983) AIChE meeting entitled "Awareness of Information Sources." The concept for this symposium evolved from discussions held after the symposium in Chicago (November 1980) entitled "Information for Chemical Engineers." Ms. Irene Farkas—Conn. was the key driving force in pointing out the need for a continuation of the series. Especially needed would be a panel discussion on information communication efforts within the Chemical Engineering profession.

The current symposium contains 16 papers and the transcription of a panel discussion on the effectiveness of information transfer. The papers are separated into three parts. Part One covers standards organizations and the Engineering Societies Library. In Part Two there are papers on data centers for material properties. Data centers for thermophysical properties are described in Part Three. The panel discussion is the final paper.

A follow up to this series is planned for Seattle (August 1985). It will be "State-of-the-Art Information Resources for Chemical Engineers."

The symposium in this volume was sponsored by The Research Committee, Group 18j in hopes that new and interesting sources of information would be open up to the profession.

Max Klein
Gas Research Institute
Chicago, Illinois

T. B. Selover, Jr.
The Standard Oil Company (OH)
Cleveland, Ohio

CONTENTS

FOREWORD III

INTERNATIONAL STANDARDS—FACTS AND FOLKLORE .... Peter G. Brown and William T. Cavanaugh ... 1

ENVIRONMENTAL MEASUREMENTS, STANDARDS, AND DECISIONS .... William H. Kirchhoff ... 6

THE ENGINEERING SOCIETIES LIBRARY: AN UNTAPPED RESOURCE FOR CHEMICAL ENGINEERS .... T.B. Selover, Jr. ... 12

THERMOPHYSICAL PROPERTIES RESEARCH AT CINDAS AND PURDUE UNIVERSITY .... Peter E. Liley ... 16

THE METAL PROPERTIES COUNCIL'S COOPERATIVE PROPERTY DATA BASE .... James A. Graham ... 25

REVIEW OF EXISTING MATERIAL PROPERTIES COMPILATIONS .... Jack H. Westbrook ... 34

ESDU—A VALIDATED ENGINEERING DATA SERVICE .... John Shaw ... 43

FACILITY FOR THE ANALYSIS OF CHEMICAL THERMODYNAMICS .... W.T. Thompson, A.D. Pelton and C.W. Bale ... 50

THERMODYNAMIC DATA GENERATION, ESTIMATION, AND COMPILATION FOR MINERAL TECHNOLOGY .... N.A. Gokcen ... 64

SOURCES OF THERMOCHEMICAL DATA—ARCHIVAL AND CURRENT .... Robert D. Freeman ... 77

AQUEOUS SOLUTIONS DATABASES AT HIGH TEMPERATURES .... Sidney L. Phillips ... 80

THERMODYNAMIC DATA FOR AQUEOUS SOLUTIONS AND THEIR USES .... Loren G. Helpler ... 84

GAS PROCESSORS ASSOCIATION—THERMOPHYSICAL DATA FOR THE GAS PROCESSING INDUSTRY .... David F. Bergman and Carl Sutton ... 93

THERMOPHYSICAL PROPERTY DATA GENERATED BY THE NBS CENTER FOR CHEMICAL ENGINEERING .... Neil A. Olien and Harold J. Raveche ... 101

THERMODYNAMIC AND THERMOCHEMICAL STUDIES AT THE BARTLESVILLE ENERGY TECHNOLOGY CENTER .... Bruce E. Gammon ... 108

THE ROLE OF THE DATA DESIGN LABORATORY .... Joseph R. Downey ... 116

DISCUSSION SECTION 123

INTERNATIONAL STANDARDS — FACTS AND FOLKLORE

Peter G. Brown and William T. Cavanaugh ■ ASTM, Philadelphia, PA

Since 1980, ASTM has intensified its study of the international use of standards with the objective of separating fact from folklore. Of particular interest in this regard, is the international use of voluntary and mandatory standards, their role in international trade, and the General Agreement on Tarriff & Trade (GATT). This paper provides an overview of the international standards system, its relationship to the national standards system, the GATT Standards Code, and highlights of a recent ASTM study of the use of standards in international trade.

Until the more recent internationalization of the marketplace, few individuals or organizations focused on the role of standards and standards organizations in international trade. Those who were involved stressed the importance of international standards, but had little data to support their views. The following summary provides an overview of the international standards system, its relationship to the national standards system, an explanation of the GATT Standards Code, and highlights a recent "ASTM Study of the Use of Standards in International Trade."

THE INTERNATIONAL STANDARDS SYSTEM

The international standards system, similar to the U. S. national standards system consists of a voluntary (private) and governmental (treaty) sector. Each of these provides an interface for nations throughout the world to meet and develop standards.

The major organizations in the voluntary international standards arena are the International Organization for Standardization (ISO) and the International Electrotechnical Commission (IEC).

The International Organization for Standardization (ISO)

Delegates from 25 countries met in London in 1946 and decided to create a new international organization "the object of which would be to facilitate the international coordination and unification of industrial standards". The new organization, ISO, began to function in 1947. Over 70 nations (member bodies) participate in ISO's 184 technical committees (2,000 technical groups). These groups have developed over 5,000 ISO standards to date. The scope of ISO covers standardization in all fields except electrical and electronic engineering standards which are the responsibility of IEC.

The International Electrotechnical Commission (IEC)

IEC came into being in 1906 and consists of approximately 43 nations that participate through National Committees, one for each country, which are required to be as representative as possible of all electrical interests in the country concerned. IEC world standards are prepared by over 200 committees and subcommittees and some 700 working groups. The aims of IEC standards are to secure communication of engineering information, the reliability and safety of equipment, inter-changeability and mutual comp.tibility of equipment and the elimination of unnecessary diversity of components used in the construction of equipment.

The treaty (governmental) organizations in the international standards arena are best exemplified by two organizations, The International Organization of Legal Metrology (OIML) and Organization for Economic Cooperation and Development (OECD).

The International Organization for Legal Metrology (OIML)

OIML was established in 1955 for the purpose of reaching international agreements on standards for measuring instruments and methods of measurement. These instruments and methods are intended for use in measuring products and commodities that are traded between nations or that may be subject to internal legal requirements. The United States joined the OIML Convention in 1972 with the Department of Commerce having the responsibility for managing U. S. participation in the organization. This responsibility was assigned to the National Bureau of Standards, an agency of the Department of Commerce. The technical activities of OIML are conducted by 31 private Secretariats, each of which has several reporting Secretariats, adding up to a total of some 200 in all. Any member country, that wishes, may participate in the work of any Secretariat.

The Organization for Economic Cooperation and Development (OECD)

The OECD standardization activities include the development of recommendations on harmonization and simplification of national regulations for specific products. The OECD has a Chemicals programme aimed at international harmonization of chemicals, control legislation and its implementation. As a treaty organization, the OECD member countries are represented by governmental agencies, the U.S. Department of State, who in turn relies on such agencies as the Environmental Protection Agency (EPA) for representation. Twenty-four countries participate in the activities of OECD on a wide range of energy, environmental and specific product standardization activities.

RELATIONSHIP OF VOLUNTARY NATIONAL STANDARDS ORGANIZATIONS TO ISO AND IEC

The U. S. is a member of ISO through the American National Standards Institute (ANSI) and a member of IEC through the U.S. National Committee (USNC), which is part of ANSI. U.S. positions in international standards activities are developed by participants from companies, technical and trade organizations, government agencies and individuals. There are approximately 20,000 participants who develop U.S. positions relating to some 2200 international standards committees.

ANSI, as the U.S. member body in ISO, holds voting membership on approximately 160 of the total 184 ISO technical committees. To provide for effective, coordinated U.S. participation in the technical work, ANSI coordinates the establishment of a U.S. Technical Advisory Group (TAG) wherever interest has been demonstrated in a particular ISO activity. Administration of the TAG is normally assigned to a trade association, technical or professional society, standards organization or government agency. ASTM administers 50 TAGS and Secretariats.

Membership on the TAG comes, in most cases, from corresponding national standards developing groups to ensure close coordination between U.S. national and international standards activities. A similar representative system is established for IEC through the U.S. national committee and its subgroups.

GATT STANDARDS CODE

In 1980, after many years of negotiations, the Agreement on Technical Barriers to Trade (popularly called the "Standards Code"), was enacted. This Agreement is designed to eliminate the use of standards-related activities as unnecessary barriers to trade. The Code entered into force on January 1, 1980. As of December 31, 1982, the Code had 36 signatories (countries). The Code seeks to establish, for the first time, international rules among governments regulating the procedures by which standards and certification systems are prepared, adopted and applied, and by which products are tested for conformity with standards and provides new international norms for signatories' national standards development policies.

The Code's provisions are applicable to all products, both agricultural and industrial. They are not applicable to standards involving services, technical specifications included in government procurement contracts, or standards established by individual companies for their own use. The Code addresses itself to governmental and non-governmental standards, both voluntary and mandatory whether developed by central governments, state and local governments and private sector organizations.

The U.S. Congress implemented the Code in the U.S. through Title IV of the Trade Agreements Act of 1979.

The provisions of the Code contain several specific obligations:

1) That standards, technical regulations and rules of certification systems should not create unnecessary obstacles to trade.

2) That countries participating in the Code are obligated to provide the same treatment to imported products as is accorded to like domestic products and to like products originating in any other country particularly in regard to standards, technical regulations, certification and testing.

3) That national or regional certification systems grant access to foreign suppliers under conditions no less favorable than those granted to domestic or member country suppliers.

4) That signatories shall use appropriate international standards, or the relevant parts, as a basis for new national standards.

5) That signatories are encouraged to specify technical regulations and standards in terms of performance rather than design or descriptive.

6) That imported products should not, in principal, have to undergo more, different or more costly tests than domestic products.

7) That signatories must use open procedures when they are developing new and/or amended mandatory standards and rules of cerification systems (publish notice of the proposed mandatory standard or rules of certification and notify the GATT Secretariat of the proposed action which in turn notifies code signatories).

8) That signatories provide copies of the proposed actions upon request; allow other signatories to make comments on these proposed actions and takes these comments into account.

National Bureau of Standards (NBS) Information Center

Title IV (on Standards) of the Trade Agreements Act of 1979, prescribes establishment of an inquiry point, a central repository for standards information, Technical Offices for agricultural and non-agricultural products and dissemination of information. The Secretary of Commerce has delegated to NBS responsibility for establishing and maintaining the U.S. Inquiry for GATT; a central repository for standards and certification information; and a Technical Office for non-agricultural products.

The NBS efforts are part of the Standards Code and Information (SCI) in the Office of Product Standards Policy. The U.S. Inquiry Point is responsible for reporting to the GATT Secretariat all proposed U.S. regulations which might significantly affect trade. It receives and maintains files of all foreign notifications issued by the GATT Secretariat, and disseminates foreign notification information through several media and directly to interested U.S. groups. Under the Code, it also serves as a focal point for inquiries about and comments on proposed regulations.

The National Center for Standards and Certification Information (NCSCI) is the national repository of standards documents. Established in 1965, NCSCI is part of NBS' SCI Program. NCSCI is designed to meet the needs of government, industry, and the public for information on standards, regulations, certification programs and related activities that affect trade and commerce. To increase effectiveness, emphasis is changing from document collection and hard copy to a more modern information processing and distribution system.

USE OF ASTM STANDARDS AND INTERNATIONAL TRADE

In 1982, ASTM and the American Society for Mechanical Engineers (ASME) funded a study on the "Use of ASTM Standards in Four European Markets". The study was conducted by the Humberside Management Center, Humberside College, (formerly Hull College), Hull, England. The study had two principal aims:

    a) to discover, from the international users' perspective the role that standards play in international trade. The awareness of and consequent implications of the GATT were to be specifically investigated.

    b) to discover the international users' reaction to ASTM standards and construct a phenomonological model associated with the purchase of ASTM standards and other publications.

Two hundred and two (202) face-to-face interviews were conducted in Great Britian, West Germany, the Netherlands, and Sweden in corporations having extensive international trade in Chemicals, Plastics and Resins, and Iron and Steel (commodities for which ASTM has extensive standardization activities).

Summary of Results[1]

The major points that were brought out by an analysis of the 202 face-to-face interviews conducted in the three commodity areas are:

1. ASTM is a major standards organization whose standards for the commodities under review are clearly *de facto* used and accepted around the world.
2. Non-domestic standards (including those from ASTM) are used extensively in Europe, many respondents claiming that these are an essential and necessary feature of international trade.
3. It is the customer who determines the choice of a particular standard to be used in a commercial transaction.
4. DIN (West Germany) is the leading European standard, having a dominant position over BSI (British Standards Institute) which is the other major indigenous standard.
5. ASTM is the dominant exogenous standard.
6. Table 4 ranks the top four standards against the major criteria that were emphasized in the study.

Table 4

Ranking of Major Standards

Rank	Users' Knowledge of Standards	Usage	Kept on Users' Premises
1	DIN	DIN	DIN
2	ASTM	BSI	ASTM
3	BSI	ASTM	BSI
4	ISO	ISO	ISO

7. A significant move to more definitive international standards was predicted, ISO being the most likely agent for their gradual introduction. However, it would be wrong to infer necessarily that these more definitive international standards will be produced only by ISO. Other standards, principally those of DIN and ASTM have achieved worldwide status by virtue of their extremely wide acceptance.
8. Promptness of standards bodies in issuing new standards, revising existing ones, and cancellation and withdrawal of obsolete standards is highly important to users.
9. There is very little awareness of the GATT Standards Code by the ultimate users of the standards.

References

[1] Use of ASTM Standards in International Trade, ASTM's Standardization News, May-June, 1984.

ENVIRONMENTAL MEASUREMENTS, STANDARDS, AND DECISIONS

William H. Kirchhoff ■ National Bureau of Standards, Washington, D.C. 20234

A pervasive and often ignored condition in the development and implementation of environmental policy is uncertainty. This paper will describe limitations of the measurement system, the role of standards in compensating for uncertainty, and features of the American Society for Testing and Materials, ASTM, and its process for setting standards which offer alternatives to litigation and confrontation as approaches to reaching consensus.

I have been asked to describe an area of ASTM activities related to the interests of chemical engineers, namely that of environmental monitoring and data. The American Society for Testing and Materials, ASTM, is the largest standards writing body in the United States. As a standards writing body, it is concerned with data, their quality and interpretability, but, with minor exceptions, it is not a source of data. Its principal products are the approximately 6,700 standards published in the 48 volumes of the Annual Book of ASTM Standards. These standards emerge through a consensus process from over 130 committees, 1900 subcommittees and uncounted task groups. ASTM activities related to environmental monitoring date as far back as 50 years, but in the last two decades the number of environmentally-related activities has grown so rapidly that a dozen committees are now related in one way or another to environmental monitoring and testing.

I do not propose to describe how the consensus process works within ASTM nor present a summary of current ASTM activities

*Contribution of the National Bureau of Standards, not subject to copyright. The opinions expressed in this document represent solely those of the author and are based on experience gained through participation in the development of standards related to environmental measurement.

in the area of environmental monitoring and data. Although such a discourse would be useful to someone contemplating participation in one of ASTM's committees, a more fundamental issue is why standard methods are needed at all. In the early stages of standards development, criticism is invariably raised by practitioners of the scientific discipline under investigation. Each time the criticisms are surprisingly the same, though each group feels its situation to be unique. One hears: "This is a research field and is not yet well enough understood to be amenable to standardization. Standardization is no guarantee of reliable or accurate data. Standardization will inhibit research and development. Standards, to be credible, can only be written by those with extensive research experience in the field, and I am much too committed to other activities to participate at this time."

It is an anomaly of the standards writing business that whereas all of these criticisms are, to varying degrees, true, the process of standardization is still vital to the application of science to practical problems. Whenever scientific and social imperatives clash, the resulting structures, devised by intelligent and well-meaning people, abound with anomalies. In understanding and accepting these anomalies, one can better appreciate the role of standards in maintaining social and scientific order.

MEASUREMENT ACCURACY

All decisions, if they are to be wisely made, rely on accurate information. The quality of a decision of a scientific or technical nature is directly dependent on the quality of its supporting data. The quality of a body of data is described in terms of two interrelated aspects, the veracity of its interpretation and the accuracy of its measurement. In the following discourse, accuracy and precision will be emphasized though the correctness of the interpretation and understanding of the data are of equal importance. We adopt here the conventional definitions of accuracy - closeness to the true value - and precision - spread of data about some average value. The concept of the true value, if investigated carefully, often results in an analysis of the measurement process itself. The true value, the measurement accuracy, and the interpretation of the measured values are fundamentally inseparable.

Different kinds of decisions require data of different quality. Striving for the maximum accuracy possible in a given situation may be a less wise strategy than considering what degree of accuracy is needed for a particular purpose or for a particular interpretation. Data without accuracy estimates are incomplete and not very useful. Sensible decisions require data of known and adequate accuracy. These are the two fundamental issues which must be addressed before the collection of data begins. What accuracy is required and how is accuracy to be assessed?

The requirements for accuracy depend upon the purposes for which the measurements are made.

Environmental monitoring is conducted, for the most part:

- to determine the state of the environment and the effect of this state on health and welfare, primarily human,

- to assess, quantitatively, the effects of human activity on the state of the environment,

- to predict, quantitatively, the effects of human activity on the state of the environment, and

- to determine compliance with environmental regulations.

Accuracy requirements can be intrinsic or extrinsic to the ultimate purpose of the measurement. In some cases, the accuracy requirements for a body of data are intrinsic to the ultimate purposes for collecting the data and are determined solely by scientific considerations. In others, the accuracy requirements are extrinsic to the ultimate purposes of the measurement, and arise from needs to satisfy social requirements. A few examples may help clarify the distinctions between these two types of requirements.

Extrinsic accuracy requirements for environmental measurements are usually associated with demonstrating compliance with environmental regulations. The requirements come not from a need to assess or to predict accurately the effects of emissions on the environment, but from the need to demonstrate compliance. Characteristically, extrinsic accuracy requirements are often high, calling for uncertainties of 10% or less. This occurs in part because the regulations themselves are infinitely precise. An emission limitation is written as 100 micrograms/hour, not 100 +/- 25 micrograms/hour. Moreover, emission standards are often set at the limit of what is believed to be achievable. Another characteristic of extrinsic accuracy requirements is that they tend to be one-sided. For example, an emission which exceeds a regulated standard but which is measured to be in compliance, is likely to have no immediate consequence and little, if any, discernable long term effects.

An emission which is within a regulated standard but which is measured as exceeding the standard will result in immediate moral and economic consequences for the violator. Driving at 60 miles per hour may not be particularly more dangerous than driving 50 miles per hour, but it is usually more expensive.

The important characteristic of extrinsic accuracy requirements is that the consequences of inaccuracy have less to do with the ultimate purpose of making the measurement than with the immediate, social consequences. Since extrinsic accuracy is not necessarily scientifically required, extrinsic accuracy requirements may be replaced, for all intents and purposes, with precision requirements.

Intrinsic accuracy requirements are typically associated with prediction of environmental states and with assessment of the effects of environmental states. The determination of the effects of halocarbon emissions on stratospheric ozone, for example, requires accurate data on the existing concentration of various atomic and molecular species in the atmosphere and on the physical and chemical properties of those species. Without accurate data, the required prediction cannot be made. Moreover, the same predictive models which require accurate data can often be used to determine just how much accuracy is required. In the example of the determination of the effects of halocarbon emissions on stratospheric ozone, such sensitivity analyses have shown that some atmospheric constituents must be known with an accuracy of 10% whereas others need only be known within a factor of two and still others within an order of magnitude.

Data obtained from measurements requiring intrinsic accuracy transcend the immediate circumstances of the measurements. They are intended for continued future use and interpretation. They are subjected to scrutiny, discussion, and review and they often find their way into compilations. On the other hand, data acquired for purposes which extrinsically require accuracy, seldom are used beyond their immediate purpose. Data collected to determine compliance with hydrocarbon emissions from refineries, will, in all likelihood, never be used in large scale predictive models of atmospheric materials balance.

Two more distinctions are worthy of note. First, individuals engaged in measurements for scientific purposes are usually not the same as those engaged in measurements for regulatory purposes. And second, intrinsic accuracy requirements not met lead to inaccurate predictions, unanswered questions and/or faulty decisions. However, if measurement methods are inadequate to meet extrinsic accuracy needs, society can usually follow an alternate course. An emission performance standard can, for example, be replaced with an emission control design standard.

To summarize, the accuracy requirements of measurements conducted for assessment and prediction are intrinsic to the scientific purposes of the data. The accuracy requirements of measurements conducted to determine compliance with environmental regulations are extrinsic to the ultimate purposes of the measurement and agreement between between parties is more important than actual accuracy.

<u>The assessment of accuracy depends upon the nature of the measurement process.</u>

Environmental measurements fall into two general categories, field and laboratory. Field measurements include epidemiology (quantification of effects and enumeration of populations), compositional analysis (how much and where), flow, and dissipation and degradation (chemical and biological). Laboratory measurements include toxicology (macro, micro, acute, chronic, and genetic) and the determination of physical and chemical properties. In general, laboratory measurements provide data for prediction whereas field measurements provide data for assessment and prediction.

Laboratory measurements are conducted on well characterized substances or systems under controlled conditions. The accuracy of laboratory measurements is assessed by a combination of approaches involving repetition, correlation, and experience. Usually, laboratory measurements are infinitely repeatable in many laboratories, using different methods. Accuracy is assumed when different methods, based on different principals of measurement, give values which fall within acceptable limits for the property being measured. The correlation of the values of properties can be used to substantiate accuracy. Similar substances are expected to exhibit similar properties. Different properties of the same substance may be related through scientific theories. In the extreme, quantitative models based on physical, chemical, and biological theories and laws can be used to test the internal self consistency of a body of data. Experience, over time and from one investigator to another, with a particular measurement method can lead to a refinement of that method such that it can be used reliably to produce data of known accuracy. The evaluation of the measurement method should be a formal one with ruggedness testing and round-robin evaluations using independently characterized systems or materials. Informal experience can aid with the assessment of quality, but mostly for the rejection of data obtained with a method known to be unreliable.

Usually, laboratory measurements are published in the peer-reviewed literature and are accompanied by estimates of their uncertainties. These estimates are made by a variety of means ranging from the judgement of the individual making the measurements to a thorough, quantitative analysis of the sources of error. Typically, laboratory data are retrievable over a prolonged period of time. This longevity coupled with the data's use and interpretation enhances the reliability of the accuracy assessment.

Field measurements are conducted on particular systems which are generally out of the control of the investigator. Ideally, the measurement should interfere as little as possible with the system being measured. Thus, measurements are made of conditions which may be changing in time and the measurements are repeatable only within certain limitations or are not repeatable at all. Field measurements are, in a sense, measurements of events.

Generally, the accuracy of field measurements cannot be assessed through repetition because variability in the system being measured and variability in the measurement process cannot be distinguished, though bounds on precision can often be estimated. Nor can the accuracy of field measurements be assessed through correlation because few theories exist which can be used to relate different measurements on actual environmental systems. At best, anomalously high or low values may be identified in a body of data, but such values cannot be rejected with certainty. Even experience gained through a thorough, formal evaluation of the measurement process will be less useful than it is with laboratory measurements because of the lack of control over the conditions of the measurement.

Accuracy of field measurements can only be assessed through formal quality assurance programs in which investigators demonstrate the quality of their data by performing, on a regular basis, the same measurements on known (though not to the investigators), controlled systems as similar as possible to the environmental system being studied.

Data acquired from field measurements are seldom published. Rather, they are frequently stored in private files, reports with limited distribution, or more recently, in computer data bases. Brevity requirements frequently preclude the presentation of information on the uncertainty in the reported data. Unlike laboratory data, field data can only be defended, never verified.

Four classes of measurements have been discussed, classes which are characterized by the nature of their purpose - scientific or social - and by the nature of the measurement itself - laboratory and field. The classification alone may not appear to be very novel or useful, but the discussion should form a basis for accepting and understanding the anomalous nature of environmental measurements and data.

THE ANOMALIES OF ENVIRONMENTAL MEASUREMENTS AND DATA

Regulations requiring compliance monitoring call for a higher degree of accuracy than is possessed by the data supporting the regulations.

Environmental regulations derive from perceived threats to health and ecological well being. Once perceived, the public demands immediate protection and recourse. For this reason, regulations are often based on uncertain, even highly uncertain data. More fundamentally, the questions of what exposures to what substances over short or prolonged periods are safe to human health may be unanswerable. Time and money are never available in sufficient quantity to provide scientifically satisfactory data. Worse yet, the pressure for action is often so great that the data supporting the eventual regulations are themselves unsupported with reliable estimates of their uncertainties. Thus it is usually impossible, in the more leisurely period following the establishment of a particular regulation, to review critically the supporting data. Moreover, with the acceptance of the actions taken, the refinement of the supporting data is usually viewed as counterproductive.

On the other hand, the regulations themselves, for reasons already discussed, frequently call for a high degree of accuracy in monitoring for compliance. The high degree of accuracy however is not compelled by scientific reasons, but through a social contract between the regulator and regulated. This suggests that agreement more than accuracy is the desired goal of compliance monitoring. Under such circumstances, it is more important that the parties involved follow some standard procedure than that such a procedure provide accurate data. Ideally, accurate data are more desirable than inaccurate data, but in compliance monitoring, the need for agreement prevails.

Data collected for demonstrating compliance with environmental regulations are better documented but less used than data supporting the corresponding regulations.

This is something of a corollary to the first anomaly. Because compliance monitoring is the fulfillment of a social contract, it must be well documented and available for public scrutiny. The documentation contains not only dates, times and values of parameters, it also references the method used for the collection of the data. The regulator and regulated recognize the immediate consequences of errors arising from uncertainty in their data, whereas the generators of data to be used as the basis for a regulation are far removed from the consequences of inaccuracy. Thus, uncertainties (for the most part imprecision) in compliance monitoring data are generally better evaluated than the uncertainties in the data supporting the regulations.

Data collected for compliance monitoring are collected by organizations unrelated to the scientific study of environmental effects. Thus, data which could be useful for the continued scientific evaluation of control measures are seldom used for that purpose.

Data acquired for use in predictive models possess greater accuracy than the predictive capability of the models and greater accuracy than the data used for the validation of the models.

This derives primarily from the nature of laboratory measurements which are inherently more accurate than field measurements. They can be verified and refined when necessary. Predictive models are simplifications of extraodinarily complex systems. At best they can predict only average values of a small set of selected parameters. Field data, collected for validation of models, are expensive to obtain and relate only to the conditions encountered when the data were collected. Field data cannot be subsequently verified.

Standards, to be useful and effective, must be developed by novices as well as experts.

This is not unique to environmental monitoring. Data for scientific studies are collected by well-trained individuals whereas the data for compliance monitoring are frequently collected by individuals with little scientific experience, little understanding of the principles underlying the measurement methods, and little appreciation of the consequences of changing conditions or steps in the measurement process. One of the benefits of the consensus development of standards is that it involves the participation of both groups of individuals. It may be the only forum in which experts and novices can debate, with mutual respect and understanding, their common experience. The involvement of novices in the development of standards is critical to their eventual effectiveness and acceptance.

Methods written solely by experts for experts will rest on assumptions of understanding and experience not available to the majority of individuals who will use those standards.

The standards development process can be more useful than the standards which are developed.

Again, this is not unique to environmental monitoring. Before the standardization process begins, little attention is paid to the comparability of data generated and reported in the scientific literature. Thoughts of standardization are usually provoked only when incompatibility of data is discovered to be embarrassingly rampant. The reasons for this are due, in part, to the notion that the precise details of measurement methods are considered too prosaic for publications which are aimed at experts in the field. The details of the measurement process are debated in ernest only when its practitioners strive to write down specifically the steps followed in the process. And it is only when the details are discussed that the sources of discrepancy become apparent. Moreover, standards development involves not only discussion but also laboratory demonstrations and intercomparisons of interim methods. By the end of the development process, the major sources of error have been uncovered and the need for the standard itself is less than when the process began.

This observation is offered in response to the true statement that standard methods do not of themselves guarantee accurate data. For that matter, certification of laboratories or individuals, which is often suggested as an alternative to the use of standard methods, does not guarantee accurate data either. (Practically speaking, laboratories or individuals are evaluated by demonstrated performance using standard methods.) Until standards are developed, however, little basis exists for assessing the degree of accuracy which can be expected to result in the collection of a body of data.

Still another benefit to be derived from the standards development process is the bringing together the regulator and regulated, prior to the time when they must face each other as regulatory adversaries, and away from the arena of litigation where skills in debate and negotiation weigh more heavily than scientific considerations.

CONCLUDING REMARKS

Without much difficulty, one could identify additional anomalies in the collection and use of environmental data. But the persistent theme to the discussion is this: When the goals of society and science seem to conflict, agreement assumes equal importance with scientific validity. Scientists cannot and should not compromise on scientific validity. But only through participation in the development of standards can they assure that methods developed are the best methods possible for the purpose at hand and can they assure that limitations in the methodology become explicit in the methods' use. Far from stifling scientific research, the discoveries resulting from the standardization process can direct future research into improved methodology. But these benefits will accrue only with the participation of the best talents possible.

Environmental regulations are inevitable and too critical to our health, the quality of our lives, and our economic well-being to be left solely in the hands of government agencies, no matter how well meaning they are. Over the past two years we have heard frequently the clarion call for regulations based on good science. But good science comes not from admonition but from participation. It is not the other fellow's responsibility, it is ours.

THE ENGINEERING SOCIETIES LIBRARY: AN UNTAPPED RESOURCE FOR CHEMICAL ENGINEERS

T.B. Selover, Jr. ■ The Standard Oil Company (Ohio), Midland Building (HB), Cleveland, OH 44115

The United Engineering Center in New York City houses national headquarters for seventeen engineering organizations. These will be discussed in their relation to the library. Library operations, scope of coverage, and services will be described in the context of how chemical engineers can make both personal and corporate use of the largest engineering library in the free world.

THE UNITED ENGINEERING CENTER

The United Engineering Center (UEC) is a unique building in New York City which is probably not even recognized by most engineers. It is the center of operations for AIChE as well as 16 other organizations. They are:

American Association of Engineering Societies (AAES)
Accreditation Board of Engineering and Technology (ABET)
American Institute of Mining, Metallurgical, and Petroleum Engineers (AIME)
American Society of Civil Engineers (ASCE)
American Society of Mechanical Engineers (ASME)
Engineering Foundation (EF)
Engineering Information, Inc. (EI)
Engineering Societies Library (ESL)
Institute of Electrical and Electronics Engineers (IEEE)
Illuminating Engineering Society (IES)
Illuminating Engineering Research Institute (IERI)
Junior Engineering Technical Society (JETS)
Metal Properties Council, Inc. (MPC)
Society of Women Engineers (SWE)
United Engineering Trustees, Inc. (UET)
Welding Research Council (WRC)

The 18 story building is located at 345 East 47th Street, New York, N.Y. 10017, one block from the United Nations. The administration of the center is carried out by the United Engineering Trustees who are responsible for operation and maintenance of the center, as well as administration and operation of the Engineering Foundation and the Engineering Societies Library.

The concept of a national engineering center had its origin in a 1904 gift by Andrew Carnegie of one and a half million dollars to erect a union building. This gift led to construction of two buildings. One was the Engineers Club. The other which opened in 1907 on West 39th Street in New York housed the engineering societies and a library.

With the passage of time the original building became too small for its occupants. This led to the construction of the present building. Former President Herbert Hoover presided at the ground breaking. The center opened for business in 1961.

By its presence as one of the major occupants of the center the Engineering Societies Library serves a unique role. It provides the cohesive force which ties together the scholarly and applied endeavors of the member engineering societies. It is also open to the public. With this introduction to the center and the role the library fullfills in its operation let us now examine the library in more detail.

THE ENGINEERING SOCIETIES LIBRARY

The volume of technical information, already increasing at a rapid rate, is likely to expand even further as engineers intensify their effort to develop the alternatives for vast quantities of petroleum resources that we now consume. In one way or another, the challenge is having an influence on the work of engineers engaged in practically every industry: heat, electricity, transportation, mining, metals, communications, construction, chemicals, food, or services. As Lavendel[1] pointed out, "With the volume of technical information exploding, a good, active, forward-looking technical library has become a matter of survival for many scientists and engineers." Stanley[2] has shown through analysis of several studies that chemical engineers are not utilizing their information centers and libraries for much of their information needs.

The Engineering Societies Library, the largest engineering library in the free world, is a major part of the answer to the dilema. The collection incorporates subject material covering all the fields of engineering. The holdings number over 275,000 volumes, 10,000 maps and more than 100 motion picture research films on fluid mechanics and heat transfer.

The library currently receives about 5,600 periodicals and serials published in more than 50 countries and in 25 languages. A list of 3000 periodicals currently received is published every two years. The current edition, published in March, 1982, covers periodicals received as of December 1981 in 270 pages. It costs $20. Journal holdings are also listed in CASSI (Chemical Abstracts Service Source Index).

A monthly selected acquisitions list averaging 250 titles is now available for a $25 annual subscription. This listing is an excellent means for engineers and librarians to identify new publications for loan, review, and subsequent purchase.

Some unusual holdings of specific interest to chemical engineers are:

A. Journals: Berichte Bunsen Gesellschaft fuer Physikalische Chemie; Chemische Technik (Leipzig); Calphad: Computer Coupling of Phase Diagrams and Thermochemistry; Cold Regions Science and Technology; Computers and Geosciences; Petroleum Chemistry (USSR); Theoretical Foundations of Chemical Engineering (USSR).

B. Encyclopedias

   1. McKetta, ed., Encyclopedia of Chemical Processing and Design, 1976 to date, Marcel Dekker, New York.

   2. Ullmanns Encyklopadie der technischen Chemie, 1972 to date, Verlag Chemie, Weinheim.

   3. Landolt - Bornstein, Springer Verlag, New York

   4. Gmelin, a complete set, Springer Verlag, New York

C. Standards: German DIN - Most complete holdings in U.S.

D. Films: Lending library of research films on fluid mechanics and heat transfer.

E. Conference Proceedings: Papers and conference proceeding in hardcopy for the seven participating engineering societies are kept permanently. Meeting preprints are not available from AIChE past 6 weeks after a meeting. The library is the only source for the societies publications. In this regard it is a priceless asset to society authors and user alike.

A member may have three books on loan at one time and may keep them up to two weeks, not counting the time in transit. A mail request should bear the member's legible signature and society affiliation. The service rate for book loans is $1.00 per week. While most of the books are available for loan, periodicals and conference proceedings are not loaned. Certain rare books and specialized reference books are loaned only after approval of the Engineering Societies Board.

Since periodical and conference publications are not available for loan outside the reading room, the library provides a copy service which includes retrieval and reproduction of the specific article. Photocopy service for regular orders ordinarily takes less than five days inhouse. Charges are:

Regular Mail Order: $5 service charge + 40¢/page
Quick Mail Order (24 hr. inhouse): $10 + 40¢/page
Phone in Quick Order (24 hr. inhouse): $15 + 40¢/page

With DIALORDER(TM) and ORBDOC(TM), both electronic order services, charges on regular and quick mail are $1 more per item. A minimum deposit account of $50 will speed processing and eliminate $1 of the service charge. Three self service coin-operated photocopy machines are available in an alcove off the reading room.

The library holdings contain nearly all materials indexed and abstracted in COMPENDEX(R) which is the computerized information retrieval system for citations in Engineering Index(R). The library can be designated as a primary supplier of documents identified in an on-line DIALOG(R) search through the DIALORDER(TM) service by using the designation ESL. This direct electronic order service began in October 1980. In October 1981, the equivalent service on ORBDOC(TM) from ORBIT(R) was added. By spring 1983 the average number of monthly on-line orders through both data bases combined reached 1295.

Literature searches done both manually and on the computerized search services are made for all purposes including that of disclosures related to patents. Therefore, all inquiries are confidential. This service ranges from listing a few references on a specific request to the preparation of comprehensive annotated bibliographies. Only those references pertinent to the inquiry are listed. The library has extensive resources of engineering publications which often make it possible to include the examination of original works. The service charge for searches is currently $35 per hour and a prorated cost for computer time. An estimate of the anticipated charges is furnished on request.

In 1982, over 145,000 requests for library services were processed. A recent random sample of 1,200 orders from among those received in one month revealed that the average length of articles copied was 10 pages. A geographic breakdown of the sources of the sample's orders follows:

Percent	Received From
21.4	N. Y. (exclusive of New York City), N.J., Pa.
17.1	New England
12.2	Mexico and overseas
11.5	Ohio, Ind. Mich. Ky.
8.7	Wash., D.C., Va., Del., S.C., N.C., Md., W.Va.
6.8	Calif., Ore., Wash., Alaska, Hawaii
6.0	Fla., Ga., Tenn., Miss., Ala.
5.4	Tex., Okla., Ark., La.
3.2	New York City
2.5	Colo., Ariz., N. M., Utah, Nev., Wyo., Idaho
2.1	Ill., Mo., Neb., Kan.
2.0	Canada
1.1	Wis., Minn., Iowa., N.D., S.D., Mont.

It is significant that only 3.2% of the requests were from the city in which the library is located.

The past year resulted in 75,000 photoprint orders for 860,000 pages copied. Reading room pages copied by users themselves totaled 258,000. Search requests, both by hand and on-line were up 22% to 209.

The library is a Department of the United Engineering Trustees, Inc. The major portion of the financial support for the library comes from two sources. Seven participating engineering societies contribute at the rate of 65 cents per member per year. Thus, the library's resources are available to the 420,000 members of the engineering community through a minimal yearly allotment of their professional society dues. The societies are:

AIChE   IEEE
ASME   SWE
ASCE   AIME
IES

The other major source of income comes from charges for services. The library's mission is a non-profit educational function. The service charges are intended to defray the specific expense of service, e.g., retrieval, the cost of photocopy, and the cost of specialized professional services such as literature searches.

The AIChE annual support mounted to 1.8% of the library revenue for 1982 through the 65 cent member subsidy. Direct library service to users in general generated 70% of the libraries operating income in 1982 while overall supporting societies assessment totaled 23%.

A recent two month sampling survey of over 3000 document requests in mail orders showed 9% were related to chemical engineering fields. The same percentage of search requests were in this field. Of the total engineering membership represented by library supporting societies, 9% are AIChE members. Thus, usage is right in line with membership support.

The library is operated by a staff of 40, including 14 professional librarians. The director is S. K. Cabeen, who is responsible for the staff and day to day operations. The library is governed by a Library Board consisting of 15 volunteers appointed by United Engineering Trustees the Founder Societies, and the Associates. AIChE representatives are H. Bieber and T. B. Selover, Jr.

The Engineering Societies Library is in its seventy first year of operation. It is located on the second floor of the United Engineering Center. Engineers can make use of their library privileges by mail or by personally visiting the reading room. Services can also be arranged by telephone during weekday business hours by calling (212) 705-7611

Except for Saturdays, Sundays and legal holidays, the hours are from 0900 to 1900 Monday through Thursday and on Friday from 0900 to 1700. In the summer, the hours are Monday through Friday from 0900 to 1700.

CONCLUSION

At $.65 per year of member dues the availability of such a complete and vital engineering information resource is truly a bargain. When in New York, make a point of visiting the library on the second floor of the United Engineering Center and checking their resources. In any event, don't forget to make use of the privileges available to AIChE members through the mail.

LITERATURE CITED

1. Lavendel, G., "Special Libraries," Chem. Eng. News, 55(16), 34(1977).

2. Stanley, W. G., "Unique Information Resources for the Chemical Engineer", Chem. Eng. Prog., 77(6), 80(1981).

THERMOPHYSICAL PROPERTIES RESEARCH AT CINDAS AND PURDUE UNIVERSITY

P.E. Liley ■ Center for Information and Numerical Data Analysis and Synthesis (CINDAS)
Purdue University, 2595 Yeager Road, West Lafayette, IN 47906

A general review is given of studies underway at the Center for Information and Numerical Analysis and Synthesis (CINDAS) and some other Purdue facilities on thermophysical (i.e., thermodynamic and transport) properties research.

In 1957 the Thermophysical Properties Research Center (TPRC) was established at Purdue University. The motivation was the awareness of a critical lack of accurate information on properties often encountered in heat transfer, a discipline which was already then very active at Purdue. The properties initially selected were thermal conductivity (including also accommodation coefficient and thermal contact resistance), shear viscosity for Newtonian fluids, specific heat at constant pressure, thermal emissivity, reflectivity, absorptivity, and transmissivity, mass diffusivity, thermal diffusivity, and the Prandtl number. It was felt convenient to use the word "thermophysical" to describe these properties which include both "equilibrium" or "thermodynamic" properties and "non-equilibrium" or "transport" properties. No restrictions were placed on the types of phase or material or on the nature of the information collected so that experimental studies of the thermal conductivity of aspirin were collected as well as theoretical estimates of the viscosity of ionized nitrogen. Different sources were used as the source of this information and an index to these, called the Retrieval Guide, was published at a somewhat later date. The information contained in the documents was then analyzed in a selective manner by staff at the Center and tables of recommended values along with a discussion of their generation and probable accuracy were generated. The original publication detailing the philosophy behind the establishment of TPRC was published in 1958 ($\underline{1}$) while the Retrieval Guide ($\underline{2}$) and the Data Books ($\underline{3}$) are still current.

After about three years of document acquisition, TPRC increased its generation of tables of recommended values and also initiated experimental measurements. In the meantime, the coverage of properties was expanded to include thermal linear and volumetric expansion and solar absorptance to hemispherical total emittance ratio. The production of literature retrieval, analytical and experimental studies continued at an increasing rate for another 13 years. Then, in 1973, the Electronic Properties Information Center (EPIC), which had been operated for 11 years at the Hughes Aircraft Co., was transferred to TPRC, and the Center was renamed CINDAS as the additional properties were no longer thermophysical. Since then, its operation has concentrated on 14 thermophysical and 22 electronic properties, and the Center is recognized as a Thermophysical and Electronic Properties Information Analysis Center (TEPIAC) by the Department of Defense.

The following sections will give a general description of typical research on these different properties at CINDAS and some related areas at Purdue University.

<u>FLUIDS</u>

The initial collection and analysis of information was on the three properties:

thermal conductivity, viscosity, and specific heat at constant pressure. The fluids selected for analysis numbered about 70, being chosen for their technological importance. They are listed in Table 1. The initial analyses were confined to the physical states: saturated liquid, saturated vapor, and the dilute (atmospheric pressure and zero pressure) gas. As described in some more detail elsewhere (4), in a number of cases the analysis of the available information leads to a few sets of experimental or theoretical values considered to be reliable, although significant discrepancies may occur between one or more of these. The resolution of these discrepancies has proven to be very time consuming, and has, at times, led to some criticism of our selection. The usual statement has been that our selection has not been sufficiently rigorous from the standpoint of theory. Figure 1 is part of a departure plot (5) for gaseous air at atmospheric pressure in which the ordinate represents the fractional deviation from an assumed base value of the property selected, in this case thermal conductivity. As stated in that publication in 1968, an estimate of the best values better than a 1% uncertainty was not then considered to be possible. For air, a multicomponent gas, one possible theoretical route would have been to use the appropriate equations for mixtures with Sutherland coefficients. However, in 1970 (6), for binary mixtures of sufficiently complex gases, accuracy better than 99%, or an uncertainty of less than 1%, was not possible. What have subsequent experimental studies shown? Scott et al. (7) state that Irving et al. (8) have obtained data which lie 1 or 2% below these values, Tsederberg and Ivanova (9) report results in good agreement in the range 0 to 100°C, and Tarzimanov and Sal'manov (10) obtained values 1 to 2% higher in this temperature range. The results of Scott et al. came 0.4 to 0.8% above these values and in all cases the accuracy (i.e., uncertainty) of the published data was said to be 2% or higher.

Since the mid-1970's, the CINDAS analysis has been extended to include the effect of pressure. As a matter of both technological needs and to limit the amount of analysis required, the upper limit of pressure considered was selected as one kilobar (10^8 N/M^2). Also, 1000 K was selected as the upper limit of temperature. Finally, the Prandtl number has been calculated from the three constituent properties. In all the tables an attempt has been made to tabulate the values for common increments of pressure and temperature. Table 2 shows our general table for air.

Our primary aim has been the provision of tabular data, even in a few cases of unknown reliability, for the immediate utilization of such values. The pressure of time has, in general, presented a detailed analysis of the physical significance of the values. The statistic that it took about two and one-half months to collect, reduce to a common system of units, to analyze, select, tabulate, and prepare departure plots for the 60 or so sources used for dilute air thermal conductivity may be instructive. Here, only a few of the more scientific studies for fluids will be mentioned. (A good summary is contained in (4).)

1. To a first approximation, assuming a linear dependence of liquid thermal conductivity with temperature, the thermal conductivity of chemically similar substances converge at one point. An example is the various values of thermal conductivity of liquid refrigerants (11) for which both the author and Missenard (12) found convergence to occur.

2. The difference between the thermal conductivities of saturated liquid and vapor at a given temperature varies linearly with the enthalpy of vaporization at the same temperature (13).

3. The semi-empirical equation,

$$\psi = \sqrt{T} / \sum_{i=0}^{3} a_i / T^i,$$

is capable of representing the dilute gas property, ψ, over ranges of several hundreds of degrees where ψ is either the thermal conductivity or the viscosity (14).

4. Lennard-Jones 6-12 potential intermolecular force parameters to a first approximation vary in a regular manner with the position of the substance in a chemical series and different series can be represented in a two-dimensional (planar) correlation (15).

5. The variation with reduced temperature of the reduced Prandtl number seems to follow a general correlation for many different liquids (16).

In all our analyses on fluids we have not considered in any detail the possibilities of energy absorption in the fluid, critical region anomalies, or decomposition/dissociation.

SOLIDS

As with the fluids, the initial Center studies were on the property thermal conductivity. One of the earliest findings was that 1002 data points for 83 samples of varying purity of 22 metals could be represented on one generalized thermal conductivity-temperature curve if account was taken of varying impurity and imperfection. Figure 2 (17) shows that the location of the maximum in thermal conductivity at cryogenic temperatures is strongly dependent upon this factor. Following this study, a large research effort resulted in the production of recommended tables of thermal conductivity for the elements and other materials (18,19), the compilations totaling 2751 pages of Data Book and 2897 pages of Handbook material. Subsequent work has considered binary alloys (20).

Other properties have been added, in part to reflect the addition of EPIC to the Center. A summary of some of this work by Ho and Touloukian (21) includes studies on electrical resistivity (22), refractive index (23), etc.

Another important group of properties is the thermal radiative properties, such as emissivity, absorptivity, reflectivity, and transmissivity. For metallic elements and alloys, 1644 pages of Handbook material have been produced (24a), for nonmetallic solids, 1890 pages (24b), and for coatings, 1569 pages (24c).

Still further work on solids includes study of specific heat of metals and alloys (24d) and nonmetals (24e), thermal diffusivity (24f), and thermal expansion (24g-h). At a more recent date efforts are concentrated to prepare volumes in a new McGraw-Hill/CINDAS Data Series on Material Properties (25), and so far four volumes have been published. Table 3 lists the proposed volumes of this series (26).

Among studies currently underway at CINDAS or recently completed are compilations of the thermophysical and electronic properties of stainless steels, nonstainless alloy and carbon steels, thermophysical properties of inorganic and organic fluids, refrigerants, etc., thermoelectric power of elements and binary alloys, and optical properties of optical materials. For further information, the Director of CINDAS may be contacted or, for the particular case of fluids, the author of this paper.

OTHER RESEARCH

As mentioned in the preceding text, initially TPRC conducted experimental as well as analytical research. On July 1, 1973 the TPRC Experimental Research Division was merged with thermophysics research activities of the School of Mechanical Engineering and administrative responsibilities were assigned to that School. Dr. R. E. Taylor of the TPRC staff was appointed Head of the Laboratory, now named the Thermophysical Properties Research Laboratory (TPRL). Care should be taken not to confuse TPRL with TPRC or with another Purdue activity, TSPC, the Thermal Sciences and Propulsion Center. TPRL is physically located in the CINDAS building where it occupies 4400 square feet of laboratory space. For a complete description of TPRL the reader is referred to TPRL 181 Rev., a 72-page brochure (27). Examples of its measurements include a multi-property apparatus which measures linear expansion, coefficient of expansion, electrical resistivity, hemispherical total emittance, thermal conductivity, specific heat capacity, normal spectral emissivity, Wiedemann-Franz-Lorenz ratio, thermal diffusivity, and the Thomson coefficient. Other equipment includes a high-temperature emissometer, modified Kohlrausch method, differential scanning calorimeter, push-rod dilatometer, thermal comparator, and heated probe apparatus. Five recent publications from TPRL are listed (28-32).

As part of the University work, students study for the Masters and Doctoral theses. Such studies can be analytical or experimental. In the past, TPRC, TPRL, and other parts of Purdue University have generated such works. Major components of the University which have contributed include the School of Mechanical Engineering itself, the Herrick Laboratories, the Thermal Sciences and Propulsion Center, and other departments such as Chemistry and Physics, etc. The author will be glad to supply further information on these activities.

LITERATURE CITED

1. Touloukian, Y.S., in *Transport Properties in Gases* (Cambel, A.B. and Fenn, J.B., Editors), Northwestern University Press, Evanston, IL, 12-24, 1958.

2. Chaney, J.F., Ramdas, V., Rodriguez, C.R., and Wu, M.H. (Editors), Thermophysical Properties Research Literature Retrieval Guide 1900-1980, Plenum Publishing Corp., New York, NY, 4962 pp., 1982.

3. Touloukian, Y.S. and Ho, C.Y. (Editors), McGraw-Hill/CINDAS Data Series on Material Properties, McGraw-Hill Book Co., New York, NY, 1981- . Earlier volumes were the Thermophysical Properties of Matter--The TPRC Data Series (Touloukian, Y.S. and Ho, C.Y., Editors), IFI/Plenum Data Co., New York, NY, 14 vols. (15 books), 16,810 pp., 1970-79.

4. Liley, P.E., in The Technological Importance of Accurate Thermophysical Property Information (Sengers, J.V. and Klein, M., Editors), NBS Rept. NBS SP 590, 43-52, 1980.

5. Liley, P.E., in Proceedings of the Fourth Symposium on Thermophysical Properties (Moszynski, J.R., Editor), ASME, New York, NY, 323-49, 1968.

6. Touloukian, Y.S., Liley, P.E., and Saxena, S.C., in Thermal Conductivity - Nonmetallic Liquids and Gases, Vol. 3 of Thermophysical Properties of Matter--The TPRC Data Series, IFI/Plenum Data Co., New York, NY, 707 pp., 1970.

7. Scott, A.C., Johns, A.I., Watson, J.T.R., and Clifford, A.A., Int. J. Thermophys., $\underline{2}$(2), 103-14, 1981.

8. Irving, J.B., Jamieson, D.T., and Paget, D.S., Trans. Inst. Chem. Eng., $\underline{51}$, 10-13, 1973.

9. Tsederberg, N.V. and Ivanova, Z.A., Teploenergetika, $\underline{18}$, 69-71, 1971; Engl. Transl.: Therm. Eng., $\underline{18}$, 100-2, 1971.

10. Tarzimanov, A.A. and Salmanov, R.S., Teplofiz. Vys. Temp., $\underline{15}$, 912, 1977; Engl. Transl.: High Temp., $\underline{15}$, 773-4, 1977.

11. ASHRAE, Thermophysical Properties of Refrigerants, New York, NY, 237 pp., 1976.

12. Missenard, A., Cahiers de la Thermique, Serie C, No. 2, Societe Francais des Thermiciens, Paris, 1971.

13. Liley, P.E., in Proceedings of the Fifth Symposium on Thermophysical Properties (Bonilla, C.F., Editor), ASME, New York, NY, 197-9, 1970.

14. The equation cited was used repeatedly in references 3, 6, and 11.

15. Bishop, P.J., "Correlations of Potential Parameters of the Lennard-Jones m-6 Potential for the Methane Derivative Refrigerant Gases," M.S. Thesis, Purdue Univ., 97 pp., 1972. See also Bishop, P.J. and Liley, P.E., in Proceedings of the Sixth Symposium on Thermophysical Properties (Liley, P.E., Editor), ASME, New York, NY, 111-16, 1973.

16. Johnson, M.W., "Correlations and Calculations of the Prandtl Number of Refrigerants," M.S. Thesis, Purdue Univ., 90 pp., 1976. See also Johnson, M.W. and Liley, P.E., in Proceedings of the Seventh Symposium on Thermophysical Properties (Cezairliyan, A., Editor), ASME, New York, NY, 739-43, 1977.

17. Slightly modified from Ho, C.Y., Powell, R.W., and Liley, P.E., J. Phys. Chem. Ref. Data, $\underline{3}$(Suppl. 1), 796 pp., 1974.

18. Touloukian, Y.S. (Editor), Thermophysical Properties Research Center Data Book, 3 volumes, Purdue University, Thermophysical Properties Research Center. Volume 1. "Metallic Elements and Their Alloys," Touloukian, Y.S., Ho, C.Y., Cezairliyan, A., Powell, R.W., DeWitt, D.P., and Buyco E.H., 1578 pp., 1960-66; Volume 3. "Nonmetallic Elements, Compounds, and Mixtures (In Solid State at NTP)," Touloukian, Y.S., Ho, C.Y., Powell, R.W., DeWitt, D.P., and Buyco, E.H., 1173 pp., 1961-66.

19. Touloukian, Y.S. and Ho, C.Y. (Editors), Thermophysical Properties of Matter - The TPRC Data Series, 14 volumes (15 books), IFI/Plenum Data Co., New York, NY, 16,810 pp., 1970-79. Volume 1. "Thermal Conductivity - Metallic Elements and Alloys," Touloukian, Y.S., Powell, R.W., Ho, C.Y., and Klemens, P.G., 1595 pp., 1970; Volume 2. "Thermal Conductivity - Nonmetallic Solids," Touloukian, Y.S., Powell, R.W., Ho,

C.Y., and Klemens, P.G., 1302 pp., 1970.

20. Ho, C.Y., Ackerman, M.W., Wu, K.Y., Oh, S.G., and Havill, T.N., J. Phys. Chem. Ref. Data, 7(3), 959-1177, 1978.

21. Ho, C.Y. and Touloukian, Y.S., in *Proceedings of the Eighth Symposium on Thermophysical Properties* (Sengers, J.V., Editor), Vol. II, ASME, New York, NY, 419-30, 1982.

22. Chi, T.C., J. Phys. Chem. Ref. Data, 8(2), 339-438, 1979; Chi, T.C., J. Phys. Chem. Ref. Data, 8(2), 439-497, 1979; Matula, R.A., J. Phys. Chem. Ref. Data, 8(4), 1147-1298, 1979; Ho, C.Y., Ackerman, M.W., Wu, K.Y., Havill, T.N., Bogaard, R.H., Matula, R.A., Oh, S.G., and James, H.M., J. Phys. Chem. Ref. Data, 12(3), 1983.

23. Li, H.H., J. Phys. Chem. Ref. Data, 5(2), 329-528, 1976; Li, H.H., J. Phys. Chem. Ref. Data, 9(1), 161-289, 1980; Li, H.H., J. Phys. Chem. Ref. Data, 9(3), 1-98, 1980.

24. Touloukian, Y.S. and Ho, C.Y. (Editors), *Thermophysical Properties of Matter - The TPRC Data Series*, 14 volumes (15 books), IFI/Plenum Data Co., New York, NY, 16,810 pp., 1970-79.

   a. Volume 7. "Thermal Radiative Properties - Metallic Elements and Alloys," Touloukian, Y.S. and DeWitt, D.P., 1644 pp., 1970.

   b. Volume 8. "Thermal Radiative Properties - Nonmetallic Solids," Touloukian, Y.S. and DeWitt, D.P., 1890 pp., 1972.

   c. Volume 9. "Thermal Radiative Properties - Coatings," Touloukian, Y.S., DeWitt, D.P., and Hernicz, R.S., 1569 pp., 1972.

   d. Volume 4. "Specific Heat - Metallic Elements and Alloys," Touloukian, Y.S. and Buyco, E.H., 830 pp., 1970.

   e. Volume 5. "Specific Heat - Nonmetallic Solids," Touloukian, Y.S. and Buyco, E.H., 1737 pp., 1970.

   f. Volume 10. "Thermal Diffusivity," Touloukian, Y.S., Powell, R.W., Ho, C.Y., and Nicolaou, M.C., 760 pp., 1973.

   g. Volume 12. "Thermal Expansion - Metallic Elements and Alloys," Touloukian, Y.S., Kirby, R.K., Taylor, R.E., and Desai, P.D., 1440 pp., 1975.

   h. Volume 13. "Thermal Expansion - Nonmetallic Solids," Touloukian, Y.S., Kirby, R.K., Taylor, R.E., and Lee, T.Y.R., 1786 pp., 1977.

25. Touloukian, Y.S. and Ho, C.Y. (Editors), *McGraw-Hill/CINDAS Data Series on Material Properties*, McGraw-Hill Book Co., New York, NY, (projected to comprise totally 42 volumes and 15,000 pages), 1981- .

   Vol. II-1. "Thermal Accommodation and Adsorption Coefficients of Gases," Saxena, S.C. and Joshi, R.K., 448 pp., 1981.

   Vol. II-2. "Physical Properties of Rocks and Minerals," Touloukian, Y.S., Judd, W.R., Roy, R.F., et al., 568 pp., 1981.

   Vol. III-1. "Properties of Selected Ferrous Alloying Elements," Touloukian, Y.S., Ho, C.Y., Bogaard, R.H., Desai, P.D., Li, H.H., et al., 285 pp., 1981.

   Vol. III-2. "Properties of Nonmetallic Fluid Elements," Liley, P.E., Makita, T., and Tanaka, Y., 224 pp., 1981.

26. Ho, C.Y., "Thermophysical and Electronic Properties Information Analysis Center (TEPIAC)," Army Materials and Mechanics Research Center Rept. AMMRC-TR-81-26, 62-3, 1981.

27. Taylor, R.E., "A Description of the Thermophysical Properties Research Laboratory," TPRL Rept. 181 (Revised), 72 pp., 1982.

28. Taylor, R.E., High Temp.-High Pressures, 11, 43-58, 1979.

29. Johnson, P.E., DeWitt, D.P., and Taylor, R.E., AIAA J., 19, 113-20, 1981.

30. Taylor, R.E., High Temp.-High Pressures, 13, 9-22, 1981.

31. Taylor, R.E., Groot, H., and Shoemaker, R.L., Space Radiative Transfer and Temperature, 83, 96-108, 1982. [AIAA Inc.]

32. Moore, R.I. and Taylor, R.E., High Temp.-High Pressures, in press.

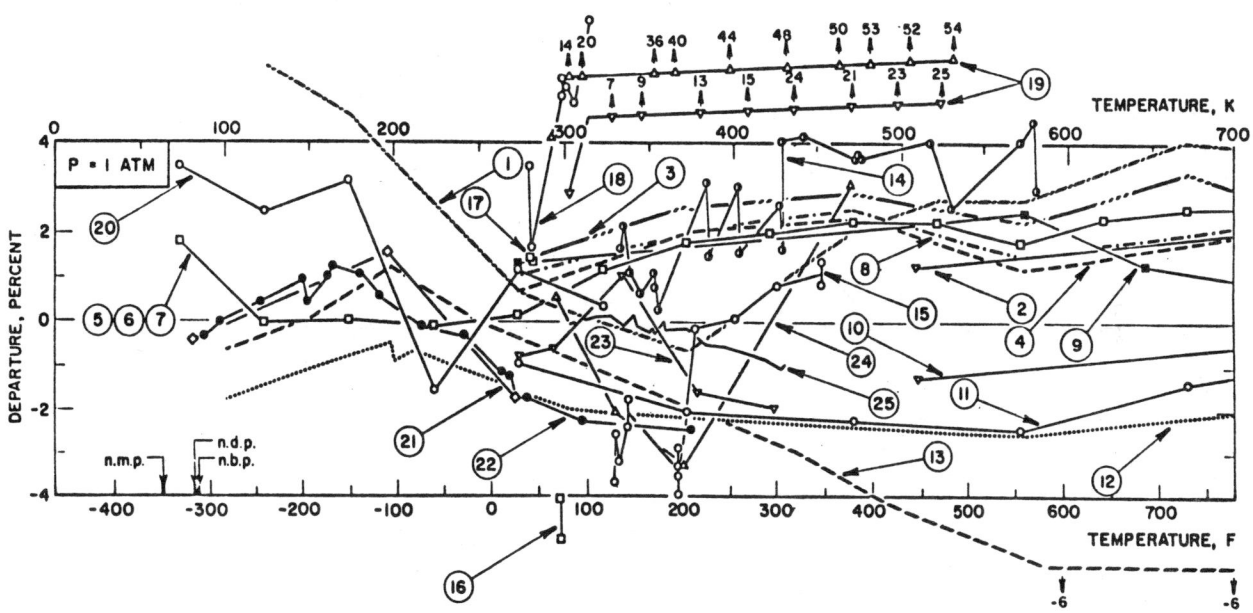

Figure 1. Departure plot for thermal conductivity of gaseous air at atmospheric pressure.

TABLE 1. FLUIDS SELECTED FOR NUMERICAL DATA ANALYSIS AT CINDAS

Acetone	Helium 4	R11 (Trichlorofluoromethane)
Acetylene	n-Heptane	R12 (Dichlorodifluoromethane)
Air	n-Hexane	R13 (Chlorotrifluoromethane)
Ammonia	Hydrogen, normal	R13B1 (Bromotrifluoromethane)
Argon	Hydrogen, para	R21 (Dichlorofluoromethane)
Benzene	Hydrogen Chloride	R22 (Chlorodifluoromethane)
Boron Trifluoride	Hydrogen Iodide	R23 (Trifluoromethane)
Bromine	Hydrogen Sulfide	R113 (Trichlorotrifluoroethane)
iso-Butane	Iodine	R114 (Dichlorotetrafluoroethane)
n-Butane	Krypton	R115 (Chloropentafluoroethane)
Carbon Dioxide	Methane	R142b (Chlorodifluoroethane)
Carbon Monoxide	Methyl Alcohol	R152a (Difluoroethane)
Carbon Tetrachloride	Methyl Chloride	R216 (1,3Dichloro1,1,2,2,3,3,-
Carbon Tetrafluoride	Neon	hexafluoropropane)
Chlorine	Nitric Oxide	R318 (Octafluorocyclobutane)
Chloroform	Nitrogen	R500 (R12,R152azeotrope)
n-Decane	Nitrogen Peroxide	R502 (R12, R115azeotrope)
Deuterium	Nitrous Oxide	R503 (R13,R23azeotrope)
Ethane	n-Nonane	R504 (R32,R115azeotrope)
Ethyl Alcohol (Ethanol)	n-Octane	Sulfur Dioxide
Ethyl Ether	Oxygen	Toluene
Ethylene	n-Pentane	Tritium
Ethylene Glycol	Propane	Water
Fluorine	Propylene	Xenon
Glycerol	Radon	

Table 2. Thermal Conductivity ($Wm^{-1} K^{-1}$) of Air as a Function of Pressure and Temperature.

Temp. (K)	1	10	20	30	40	50	60	80	100	150	200	250	300	350	400	450	500	600	700	800	900	1000
60	0.180	0.181	0.181																			
65	0.171	0.172	0.173	0.173	0.174	0.175																
70	0.163	0.164	0.165	0.165	0.166	0.167	0.168	0.169														
75	0.154	0.155	0.156	0.156	0.157	0.158	0.159	0.160	0.162	0.165	0.169											
80	0.146	0.147	0.148	0.149	0.150	0.150	0.152	0.154	0.157	0.161	0.165	0.168										
85	0.0079	0.137	0.138	0.139	0.140	0.141	0.142	0.144	0.145	0.150	0.154	0.157	0.161	0.164	0.168							
90	0.0083	0.128	0.130	0.131	0.132	0.133	0.134	0.136	0.137	0.142	0.146	0.150	0.154	0.158	0.161	0.164	0.168					
95	0.0087	0.120	0.121	0.122	0.123	0.124	0.125	0.128	0.130	0.134	0.139	0.143	0.147	0.151	0.155	0.158	0.162					
100	0.0092	0.111	0.112	0.114	0.115	0.116	0.117	0.120	0.122	0.127	0.132	0.137	0.141	0.145	0.149	0.152	0.156					
105	0.0097	0.102	0.104	0.105	0.107	0.108	0.110	0.112	0.115	0.120	0.125	0.130	0.135	0.139	0.143	0.147	0.150					
110	0.0101	0.0120	0.095	0.097	0.098	0.100	0.102	0.105	0.107	0.114	0.119	0.124	0.129	0.133	0.137	0.141	0.145					
115	0.0106	0.0123	0.086	0.088	0.090	0.092	0.094	0.097	0.100	0.107	0.113	0.118	0.123	0.128	0.132	0.136	0.140					
120	0.0111	0.0126	0.0152	0.079	0.081	0.084	0.086	0.090	0.094	0.101	0.107	0.113	0.118	0.123	0.127	0.131	0.135					
125	0.0115	0.0128	0.0152	0.066	0.072	0.075	0.078	0.083	0.087	0.095	0.102	0.108	0.113	0.118	0.123	0.127	0.131					
130	0.0120	0.0133	0.0152	0.0180	0.059	0.065	0.069	0.076	0.080	0.089	0.097	0.103	0.109	0.114	0.118	0.123	0.127					
140	0.0129	0.0141	0.0157	0.0181			0.060	0.067		0.079	0.087	0.094	0.100	0.105	0.110	0.114	0.119					
150	0.0138	0.0149	0.0162	0.0179						0.068	0.078	0.085	0.091	0.097	0.102	0.107	0.111					
160	0.0146	0.0157	0.0169	0.0183	0.0201					0.059	0.069	0.077	0.084	0.089	0.095	0.099	0.104					
170	0.0155	0.0165	0.0175	0.0188	0.0203	0.0220				0.051	0.062	0.070	0.077	0.083	0.088	0.093	0.097					
180	0.0164	0.0173	0.0182	0.0194	0.0206	0.0221	0.0238	0.028	0.033	0.046	0.056	0.064	0.071	0.077	0.082	0.087	0.091					
190	0.0172	0.0181	0.0190	0.0200	0.0211	0.0224	0.0238	0.0271	0.031	0.042	0.051	0.059	0.066	0.072	0.077	0.082	0.087					
200	0.0181	0.0189	0.0197	0.0206	0.0217	0.0228	0.0240	0.0268	0.0300	0.039	0.048	0.055	0.062	0.068	0.073	0.078	0.082	0.090				
210	0.0189	0.0197	0.0205	0.0213	0.0223	0.0233	0.0243	0.0266	0.0295	0.037	0.045	0.052	0.058	0.064	0.069	0.074	0.078	0.086	0.093			
220	0.0198	0.0205	0.0212	0.0220	0.0229	0.0238	0.0248	0.0269	0.0293	0.0360	0.0429	0.049	0.056	0.061	0.066	0.071	0.075	0.083	0.090	0.096		
230	0.0206	0.0213	0.0220	0.0228	0.0236	0.0244	0.0253	0.0272	0.0294	0.0353	0.0415	0.0476	0.053	0.058	0.063	0.068	0.072	0.080	0.087	0.093	0.099	
240	0.0214	0.0221	0.0228	0.0235	0.0242	0.0250	0.0258	0.0276	0.0295	0.0348	0.0405	0.0461	0.0514	0.056	0.060	0.065	0.069	0.077	0.084	0.090	0.096	0.100
250	0.0222	0.0229	0.0235	0.0242	0.0248	0.0259	0.0264	0.0281	0.0298	0.0347	0.0398	0.0449	0.0499	0.0546	0.059	0.063	0.067	0.075	0.081	0.087	0.093	0.099
260	0.0230	0.0237	0.0243	0.0249	0.0256	0.0263	0.0270	0.0286	0.0302	0.0346	0.0393	0.0441	0.0487	0.0532	0.0574	0.0614	0.0653	0.072	0.079	0.085	0.090	0.096
270	0.0239	0.0244	0.0250	0.0256	0.0263	0.0269	0.0276	0.0291	0.0306	0.0347	0.0391	0.0435	0.0478	0.0520	0.0561	0.0599	0.0636	0.0705	0.077	0.083	0.088	0.094
280	0.0246	0.0251	0.0257	0.0263	0.0269	0.0275	0.0282	0.0296	0.0311	0.0349	0.0389	0.0431	0.0471	0.0511	0.0550	0.0586	0.0622	0.0689	0.0751	0.081	0.086	0.092
290	0.0254	0.0259	0.0264	0.0270	0.0276	0.0282	0.0289	0.0302	0.0315	0.0351	0.0389	0.0428	0.0466	0.0504	0.0540	0.0576	0.0610	0.0674	0.0735	0.0792	0.085	0.090
300	0.0261	0.0266	0.0271	0.0277	0.0283	0.0289	0.0295	0.0307	0.0320	0.0354	0.0390	0.0426	0.0462	0.0498	0.0533	0.0567	0.0599	0.0662	0.0720	0.0776	0.0828	0.0878
320	0.0276	0.0281	0.0286	0.0291	0.0296	0.0302	0.0307	0.0320	0.0331	0.0361	0.0393	0.0425	0.0458	0.0490	0.0522	0.0553	0.0583	0.0641	0.0697	0.0749	0.0799	0.0847
340	0.0290	0.0294	0.0299	0.0304	0.0309	0.0315	0.0320	0.0331	0.0342	0.0369	0.0398	0.0427	0.0457	0.0486	0.0515	0.0544	0.0572	0.0626	0.0678	0.0727	0.0775	0.0821
360	0.0304	0.0307	0.0306	0.0311	0.0316	0.0322	0.0327	0.0337	0.0347	0.0373	0.0401	0.0429	0.0458	0.0486	0.0512	0.0540	0.0567	0.0617	0.0665	0.0711	0.0755	0.0799
380	0.0317	0.0314	0.0319	0.0324	0.0332	0.0335	0.0340	0.0350	0.0360	0.0383	0.0411	0.0436	0.0461	0.0487	0.0511	0.0539	0.0563	0.0610	0.0655	0.0700	0.0741	0.0782
400	0.0331	0.0333	0.0339	0.0343	0.0347	0.0352	0.0357	0.0368	0.0376	0.0398	0.0420	0.0443	0.0466	0.0489	0.0512	0.0535	0.0557	0.0602	0.0645	0.0688	0.0728	0.0768
450	0.0363	0.0366	0.0372	0.0375	0.0379	0.0384	0.0390	0.0398	0.0405	0.0424	0.0423	0.0462	0.0481	0.0501	0.0521	0.0540	0.0560	0.0598	0.0636	0.0673	0.0709	0.0744
500	0.0396	0.0399	0.0403	0.0408	0.0412	0.0416	0.0420	0.0428	0.0434	0.0451	0.0467	0.0484	0.0501	0.0518	0.0535	0.0552	0.0569	0.0603	0.0630	0.0669	0.0701	0.0733
600	0.0456	0.0460	0.0463	0.0467	0.0471	0.0475	0.0478	0.0458	0.0463	0.0478	0.0517	0.0531	0.0544	0.0588	0.0571	0.0585	0.0598	0.0625	0.0652	0.0679	0.0705	0.0732
700	0.0513	0.0516	0.0520	0.0524	0.0528	0.0532	0.0535	0.0541	0.0545	0.0556	0.0567	0.0578	0.0589	0.0600	0.0611	0.0623	0.0634	0.0656	0.0679	0.0701	0.0724	0.0746
800	0.0570	0.0572	0.0575	0.0578	0.0581	0.0584	0.0587	0.0592	0.0595	0.0605	0.0614	0.0624	0.0634	0.0643	0.0653	0.0662	0.0672	0.0691	0.0710	0.0730	0.0749	0.0768
900	0.0624	0.0627	0.0630	0.0632	0.0634	0.0636	0.0638	0.0642	0.0645	0.0654	0.0662	0.0671	0.0679	0.0687	0.0696	0.0704	0.0712	0.0730	0.0746	0.0763	0.0780	0.0797
1000	0.0673	0.0675	0.0678	0.0681	0.0684	0.0687	0.0690	0.0693	0.0696	0.0703	0.0711	0.0718	0.0726	0.0733	0.0740	0.0748	0.0755	0.0770	0.0785	0.0800	0.0815	0.0830

TABLE 3. STRUCTURE AND SCOPE OF "McGRAW-HILL/CINDAS DATA SERIES ON MATERIAL PROPERTIES

Group I. Theory, Estimation, and Measurement of Properties

  Vol. I-1. Transport Properties of Fluids: Thermal Conductivity, Viscosity, and Diffusion Coefficient
  Vol. I-2. Transport Properties of Solids: Thermal Conductivity, Electrical Resistivity, and Thermoelectric Properties
  Vol. I-3. Specific Heat of Solids
  Vol. I-4. Thermal Expansion of Solids
  Vol. I-5. Thermal Radiative Properties of Solids

Group II. Properties of Special Materials

  Vol. II-1. Thermal Accommodation and Adsorption Coefficients of Gases
  Vol. II-2. Physical Properties of Rocks and Minerals
  Vol. II-3. Optical Properties of Optical Materials
  Vol. II-4. Thermal Radiative Properties of Coatings

Group III. Properties of the Elements

  Vol. III-1. Properties of Selected Ferrous Alloying Elements (Cr, Co, Fe, Mn, Ni, and V)
  Vol. III-2. Properties of Nonmetallic Fluid Elements (Ar, Br, Cl, F, He, H_2, I, Kr, Ne, N_2, O_2, Rn, and Xe)
  Vol. III-3. Properties of Selected Refractory Elements (Hf, Mo, Nb, Ta, Ti, W, and Zr)
  Vol. III-4. Properties of Liquid Metal Elements (Li, Na, K, Rb, Cs, Fr, Hg, Ga, and In)
  Vol. III-5. Properties of Selected Nonferrous Alloying Elements and Precious Metals (Al, Be, Cd, Cu, Pb, Mg, Sn, Zn, Au, Ir, Pd, Pt, Re, Rh, and Ag)
  Vol. III-6. Properties of Rare-Earth and Radioactive Elements (Sc, Y, La, Ce, Pr, Nd, Pm, Sm, Eu, Gd, Tb, Dy, Ho, Er, Tm, Yb, Lu, Tc, Po, At, Rn, Fr, Ra, Ac, Th, Pa, U, Np, Pu, Am, Cm, Bk, Cf, Es, Fm, Md, No, and Lw)
  Vol. III-7. Properties of Selected Semiconducting, Semimetallic, Nonmetallic Solid, and Other Elements (Ge, Po, Se, Si, Te, Sb, As, Bi, At, B, C, P, S, Ba, Ca, Os, Ru, Sr, and Tl)

Group IV. Properties of Alloys and Cermets

  Vol. IV-1. Properties of Stainless Steels
  Vol. IV-2. Properties of Nonstainless Alloy Steels, Carbon Steels, and Cast Irons
  Vol. IV-3. Properties of Selected Transition-Metal Alloys (Alloys of Cr, Co, Hf, Mn, Mo, Ni, Nb, Pd, Pt, Rh, Ta, Ti, W, U, V, and Zr)
  Vol. IV-4. Properties of Selected Nontransition-Metal Alloys (Alloys of Al, Sb, Be, Bi, Cd, In, Pb, Mg, Sn, and Zn)
  Vol. IV-5. Properties of Copper Alloys, Gold Alloys, and Silver Alloys
  Vol. IV-6. Properties of Cermets

Group V. Properties of Fluids and Fluid Mixtures

  Vol. V-1. Properties of Inorganic and Organic Fluids
  Vol. V-2. Properties of Commercial Refrigerants and Fluid Mixtures

Group VI. Properties of Oxides and Oxide Mixtures

  Vol. VI-1. Properties of Rare-Earth Oxides and Actinide Oxides (Oxides of Sc, Y, La, Ce, Pr, Nd, Pm, Sm, Eu, Gd, Tb, Dy, Ho, Er, Tm, Yb, Lu, Ac, Th, Pa, U, Np, Pu, and Am)
  Vol. VI-2. Properties of Electronic Oxides (Oxides of Cr, Co, Cu, Fe, Mn, Ni, Ti, V, and Zn)
  Vol. VI-3. Properties of Selected Nontransition-Metal Oxides (Oxides of Al, Sb, Ba, Be, Bi, Cd, Ca, Cs, Fr, Ga, Ge, Au, In, Pb, Li, Mg, Hg, Po, K, Ra, Rb, Ag, Na, Sr, Tl, and Sn)
  Vol. VI-4. Properties of Selected Transition-Metal Oxides and Oxides of Selected Nonmetallic Solid Elements (Oxides of Hf, Ir, Mo, Nb, Os, Pd, Pt, Re, Rh, Ru, Ta, Tc, W, Zr, As, B, P, Se, Si, and Te)

TABLE 3. (CONTINUED)

Vol. VI-5. Properties of Complex Oxides
Vol. VI-6. Properties of Oxide Mixtures
Vol. VI-7. Properties of Ceramics and Glasses

Group VII. Properties of Commercial Graphites, Composites, and Systems

Vol. VII-1. Properties of Commercial Graphites and Carbon-Carbon Composites
Vol. VII-2. Properties of Composites (Other than Carbon-Carbon Composites)
Vol. VII-3. Properties of Systems

Group VIII. Properties of Non-Oxide Inorganic Compounds and Intermetallic Compounds

Vol. VIII-1. Properties of Halides (Bromides, Chlorides, Fluorides, and Iodides)
Vol. VIII-2. Properties of Borides, Carbides, Hydrides, Nitrides, and Silicides
Vol. VIII-3. Properties of Arsenides, Phosphides, Selenides, Sulfides, and Tellurides
Vol. VIII-4. Properties of Carbonates, Nitrates, Phosphates, Silicates, and Sulfates
Vol. VIII-5. Properties of Intermetallic Compounds

Group IX. Properties of Polymers, Organic Compounds, Foods, Biological Materials, and Building Materials

Vol. IX-1. Properties of Polymers
Vol. IX-2. Properties of Organic Compounds, Foods, and Biological Materials
Vol. IX-3. Properties of Building Materials

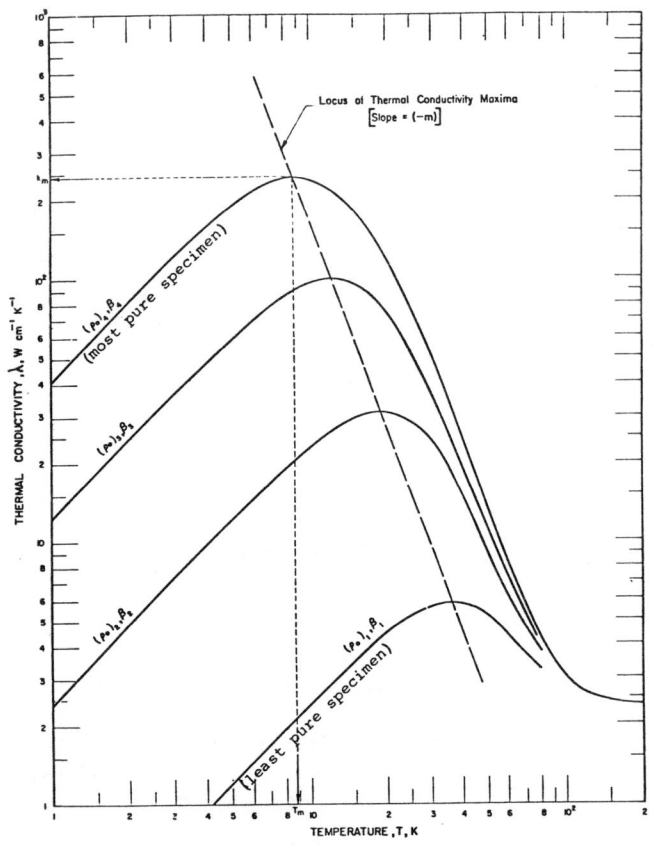

Figure 2. Low-temperature thermal conductivity of different samples of a metallic element with different impurity and imperfection content.

THE METAL PROPERTIES COUNCIL ACTIVITIES ON A NATIONAL MATERIALS PROPERTY DATA NETWORK

J.A. Graham ■ Deere & Company, Moline, Illinois

The MPC is leading an effort to establish a National Cooperative Materials Property Data Network to provide on-line access to the wealth of existing and future numerical data on the engineering properties of materials. This system would allow the user automatic, immediate access to whichever data banks in the Cooperative contain the information of interest. The data would be evaluated in terms of standards established for the Cooperative and appropriately indicated to the user.

The Cooperative would be composed of a large number of autonomous computerized data centers. The maintainers of the data centers would be technical societies, private companies, universities, research institutes and government agencies.

The paper presents some of the past activities, the current status, a statement of phased objectives and the plans to bring the data network into operation.

GOALS

Access to a Materials Properties Data Base has been desired for years by many engineers. Many developments in computers, software for data base management systems, CAD/CAM and materials properties measurement have increased the demand for and made such a system possible. The Metal Properties Council is now engaged in an effort to bring about a National Cooperative Materials Property Data Network (NC MP DN). After several years of study, funding is now being obtained to establish the activity. The goals of this project are:

- To set up in three to five years a comprehensive working materials property system to be used in certain specific application areas.
- To expand in four to eight years the initial working system into a broad scope materials information system to meet the diverse needs of the entire materials community.
- To develop the above system by a cooperative effort of industry, technical societies, government, and academia.

This is a large task. Much needs to be done, and the cooperation of many organizations is required. The measures of success for this activity will be:

- The network becomes the materials information source of first resort for information seekers and recorders.
- People use it enough so it becomes economically self-supporting.

This means having high quality, evaluated data for use in critical applications as well as general data for noncritical applications.

MANAGEMENT

A special Task Group on Computerized Data Storage and Evaluation of MPC will serve as the nucleus for technical support to a founding Board of Governors for the Data Network. A not-for-profit autonomous corporation will be established with MPC acting as secretariat. The corporation will be organized with a Board of Governors and a Chairman. It will include representatives of major industries (aerospace, aluminum, automotive, chemical, off-road vehicles, petroleum, steel, etc.) and sponsoring organizations and agencies. An Executive Director will be employed to carry out the organizational plan, build the necessary staff, establish coordination of the Independent Data Centers, and develop the network of "on-line" systems which answer current engineering needs. The initial system will aim to incorporate existing data bases which are not yet coordinated nor fully evaluated.

THE DISTRIBUTED DATA BASE

Following the recommendation of the Numerical Data Advisory Board Panel on Mechanical Properties of Metals and Alloys (1980)[1], a plan for a distributed data base network was chosen. This permits each data base developer to maintain control of his own data. Other studies which have arrived at similar conclusions are:

- The Metal Properties Council Feasibility Study by Jack Westbrook (1982)[2]
- Computerized Materials Data Workshop, Fairfield Glade, Tennessee (November 1982)[3]
- The National Materials Advisory Board of the National Research Council Panel on Computerized Mechanical Properties Data for use in CAD/CAM (1981-1983)[4]

The Workshop "Towards a National Science and Technology Data Policy," Library of Congress, April 14, 1982, coordinated by Committee on Science and Technology, U. S. House of Representatives; Congressional Research Service, Library of Congress; Numerical Data Advisory Board, National Academy of Sciences, documented a plea by major industry for evaluated data with the indication that no industry alone could take on such a project, but required the existence of reliable data resources for the conduct of business.

According to the current plan, the users would dial one phone number from their terminals and be connected to a master control computer. This concept is shown in Figure 1. After a few initial keystrokes in which the user indicates the nature of the inquiry, the control computer will tell the user what data bases are available that have information on the materials and properties they are seeking. The user then chooses the data bases he desires to search. The master computer will have the software capability to interface and search the computers of the data base suppliers, and then return the desired information to the user. Once the user has looked at the data and desires to proceed with mathematical analysis, plotting, or printout, he can use the analytical tools resident on the master computer or transfer the data to a local smart terminal that has similar tools.

The advantage of the distributed network is that each data supplier will have control over and responsibility for their data cost and operation while at the same time users will have access to data.

DATA INPUT

A data base to serve both the critical uses as well as general applications must contain several levels of data. Evaluators and certain users of data will want to look at each test value so the **unwashed** test data must be included. Other users will want

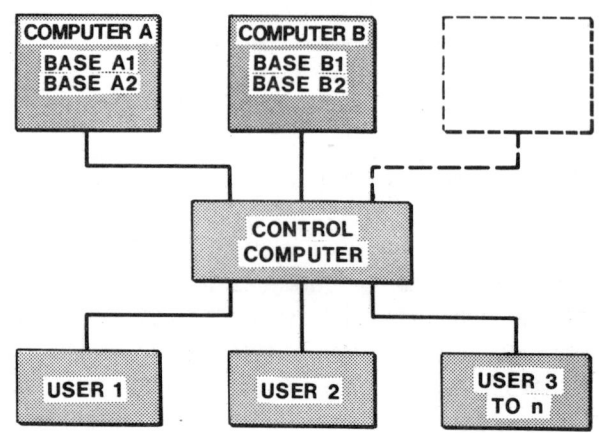

FIGURE 1. CONCEPT OF DISTRIBUTED NETWORK

to look at values that have been arrived at by an evaluator and will be looking for typical values, averages, ranges and confidence levels or recommended property values which we will call **resolved data**. The process of evaluation of data will be discussed later.

The content of the input information is extremely important or the test data will be meaningless. All of the input information remains to be determined at this time and steps have been taken to get user recommendations through a series of workshops. However, some of the input needs already identified are:

- Material designation
- Characteristics of material tested
- Test method
- Test conditions
- Test equipment
- Test results
- Reference to original source document

In each of these areas we must establish the computer input. Appropriate committees will be formed to agree on this with the aid of at least three workshops for different industries.

MATERIAL DESIGNATION

Several designations of materials are in use such as the: Unified Numbering System, trade associations, technical societies, company, government and other country designations. Which should be used in the data network? We need a designation concept that aids the use of the data network instead of one that hinders use. Such a system must permit both the user and data base supplier to use

FIGURE 2. MATERIAL DATA SUMMARY

MATERIAL FILE NAME: A22HL
LAB. NUMBER: 8944, 8945
GENERIC MATERIAL NAME:
MATERIAL NAME: A22H
SPECIMEN CONFIGURATION: ROUND AND POLISHED, 6 mm TEST DIA.
SUPPLIER:
SUPPLIER THERMO-MECHANICAL HISTORY: CD

CHEMISTRY:

	A	B	C
C	.23	.21	
Mn	.80	.80	
P	.02	.02	
S	.01	.01	
Si	.02	.01	
Ni	.01	.01	
Cr	.02	.02	
Mo	.03	.03	
Cu	.04	.04	

METALLURGY:

HARDNESS, ROCKWELL B

	A	B	C
NEAR UPPER	85	84-85	
NEAR LOWER	88-89	86-87	
CENTER	83-85	83-85	

Gage length of the test specimens will be from the center portion only.

A. GENERAL

Both tubes have a microstructure of low carbon intermediate transformation products and ferrite. The amount of ferrite varies from 25 to 40 percent in the surface areas and from 40 to 50 percent in the center portion.

B. SURFACE DECARBURIZATION

Both tubes have partial decarbonization at the surface to a depth of approximately 0.1 mm.

C. AS-RECEIVED GRAIN SIZE

Grain boundary ferrite in areas having the least ferrite indicates both tubes have an as-received grain size of 3 to 4.

D. INCLUSIONS (ASTM DESIGNATION E45, METHOD A, PLATE I)

	A	B	C
TYPES A AND C			
THIN	3	3-4	
HEAVY	1-2	2	

Although several inclusions resemble type B (alumina), EDS analysis disclosed the inclusions are all sulfides (type A) and silicates (type C).

The inclusion contents are considerable, especially in sample B, and should have a significant adverse effect on transverse properties.

their normal designation for materials, but system software must contain the ability for cross reference to other organization designations. Abbreviations, definitions, dictionary and thesaurus must be automatically available on the screen at the user's request.

CHARACTERISTICS OF THE MATERIALS TESTED

Many factors contribute to the properties of a given material sample. The most significant factors must be included to aid in interpreting the properties measured. Some of these are:

- Chemistry
- Heat treatment
- Product form
- Product dimensions
- Heat No. – Source
- Mechanical history
- Grain size
- Microstructure

Examples of data sheets used by some organizations are shown in Figures 2 through 5.

TEST PROCEDURES

Since the mechanical properties are so dependent upon the test procedure, information must be supplied to permit proper use of test results. Some of the factors to be included are:

- Specimen type
- Specimen size
- Specimen orientation
- Test method
- Test equipment
- Test conditions
- Test machine calibration
- Control of test

Again, the recommendations of a committee are needed following the workshops.

TEST RESULTS

Evaluators of the data will wish to see every data point from the tests. Examples of raw data reporting are shown in Figures 6 and 7. The "Material File Name" is the file designation where all information on this data set are located, and it is needed for both input and recall of the data.

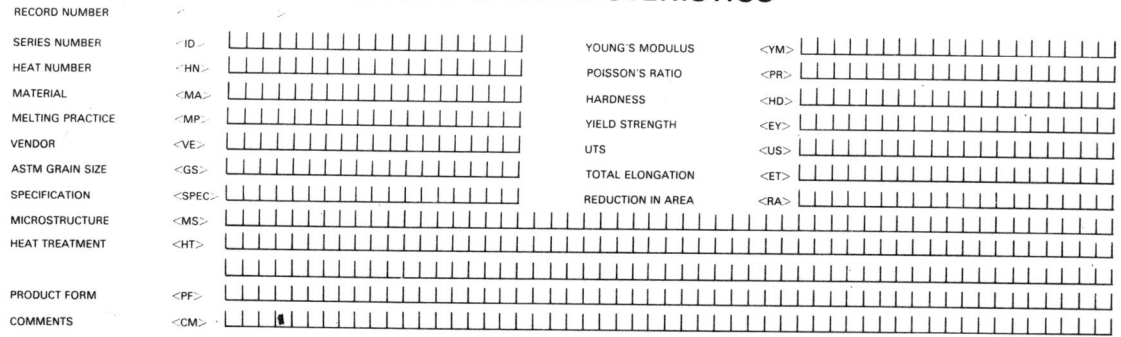

FIGURE 3. MECHANICAL PROPERTIES DATA STORAGE FORM
MATERIAL CHARACTERISTICS

FIGURE 4. MECHANICAL PROPERTIES DATA STORAGE FORM
IMPACT-SLOW BEND

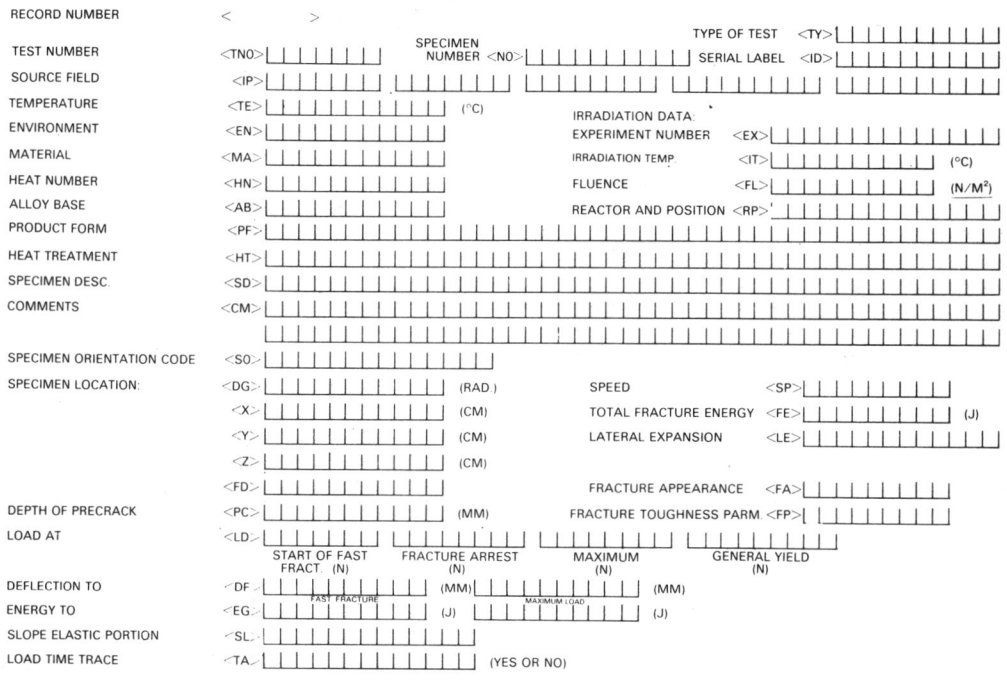

FIGURE 5. MECHANICAL PROPERTIES DATA STORAGE FORM
FATIGUE

FIGURE 6. MONOTONIC DATA

```
RAW DATA TABLE

MATERIAL FILE NAME    S1045A133
DATE ANALYSIS DONE   WED, JUN 29 1983
MATERIAL NAME:    STEEL, MED CARBON
MATERIAL GRADE NUMBER:   1045
OTHER SPECIFICATIONS FOR MATERIAL:    SAE 1045
MATERIAL CONDITION:     NORMALIZED
DATE DATA ENTERED:     THU, APR 07 1983
Bhn:   148/157
PHYSICAL ORIENTATION:    LONGITUDINAL

MONOTONIC MODULUS, E
0.203350E+06   0.201400E+06   0.000000E+00   0.000000E+00   0.000000E+00
YIELD STRENGTH, SY
0.384000E+03   0.379000E+03   0.000000E+00   0.000000E+00   0.000000E+00
ULTIMATE STRENGTH, SU
0.000000E+00   0.000000E+00   0.000000E+00   0.000000E+00   0.000000E+00
STRENGTH COEFFICIENT, K
0.119700E+04   0.117200E+04   0.000000E+00   0.000000E+00   0.000000E+00
STRAIN HARDENING EXPONENT, n
0.230000E+00   0.230000E+00   0.000000E+00   0.000000E+00   0.000000E+00
% REDUCTION IN AREA, %RA
0.510000E+02   0.510000E+02   0.000000E+00   0.000000E+00   0.000000E+00
TRUE FRACTURE STRENGTH
0.992000E+03   0.977000E+03   0.000000E+00   0.000000E+00   0.000000E+00
TRUE FRACTURE DUCTILITY
0.710000E+00   0.710000E+00   0.000000E+00   0.000000E+00   0.000000E+00
```

FIGURE 7. FATIGUE DATA

```
RAW DATA TABLE

MATERIAL FILE NAME    S1045A133
DATE ANALYSIS DONE   WED, JUN 29 1983

STRAIN    STRESS    PLASTIC    FATIGUE    CYCLE    REF
AMPL      AMPL      STRAIN     LIFE       MOD      #
                    AMPL       REV        ELAST

0.020000   524.00   0.017411       514.   202375.   133.00
0.015000   499.00   0.012534       770.   202375.   133.00
0.010000   452.00   0.007767      3054.   202375.   133.00
0.010000   465.00   0.007702      2922.   202375.   133.00
0.008000   445.00   0.005801      6088.   202375.   133.00
0.008000   440.00   0.005826      4093.   202375.   133.00
0.006000   400.00   0.004023     13344.   202375.   133.00
0.006000   420.00   0.003925     13650.   202375.   133.00
0.005000   372.00   0.003162     25826.   202375.   133.00
0.004000   351.00   0.002266     35970.   202375.   133.00
0.004000   353.00   0.002256     40398.   202375.   133.00
0.003000   315.00   0.001443     73860.   202375.   133.00
0.002500   298.00   0.001027    234268.   202375.   133.00
0.002000   270.00   0.000666    523222.   202375.   133.00
0.002000   269.00   0.000671    762902.   202375.   133.00
0.001500   241.00   0.000309   4901750.   202375.   133.00
```

REFERENCES

References are needed to aid people in going to original sources of test results for any clarification of test results.

EVALUATION

Many companies develop expertise in a limited number of materials, even though they have use for a much larger number of materials. On those materials where a company or person is an expert, they may wish to look at all the data in all the bases in the system and compare it to their own test results, thus further enhancing their knowledge. Effort is presently being devoted to arriving at the best methods of combining data sets for comparison. We know that simply putting all data together is not the answer. An example is shown in Figure 8. Drawing one line through all the data is misleading.

In the analysis shown, the two sets of data have the same slope but different ordinates. Analyzing each data set separately and then combining the results appear to be the proper method.

FIGURE 8. COMBINING DATA SETS VS DATA POINTS

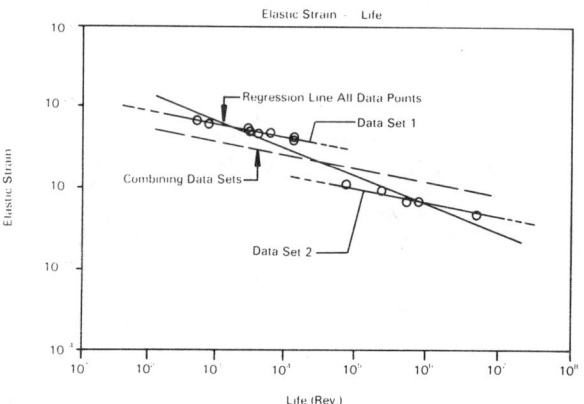

A user wanting some minimum quality data for a noncritical application may access the data for a few observations or use the resolved information. In between these two extremes the user would have the option of looking at as much or little data as he desired. Users could also perform rigorous statistical analysis or plot the data in a number of ways. Those organizations with critical applications will no doubt choose to critically evaluate each test observation, compare several sets of data from different researchers, perform their own statistical analysis, and draw conclusions as to the design values for each application. Code-setting bodies will also want to follow this procedure. By having the on-line system, more data can be reviewed and compared to make these critical decisions than by former hand methods.

DATA QUALITY

Users ask for the quality of the data. A means must be developed to provide the users on their terminals some factor(s) that show the quality. The quality designation should include at least the following factors:

- The accuracy and precision of the test results
- The degree of evaluation performed on the data

OUTPUT

The output may take several forms including disks or tapes of selected data, hard copy, tabulations, tables, graphs, or equations. In order to serve the needs of the large community of users, the data must be available for people without terminals as well as to those with extensive computer capabilities. Examples of hard copy output from one company data base are shown in Figures 9, 10, 11, and 12.

FIGURE 9. GRAPHICAL OUTPUT

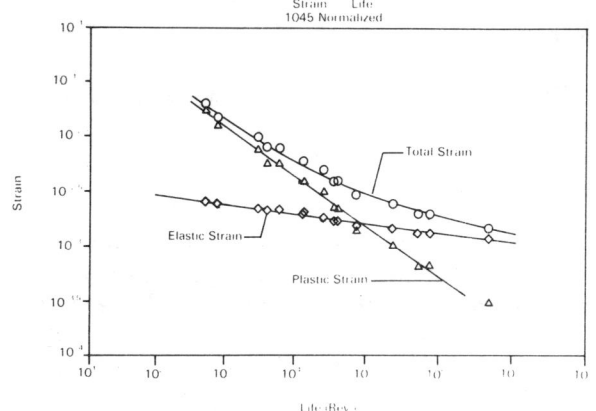

FIGURE 10. EQUATION OUTPUT

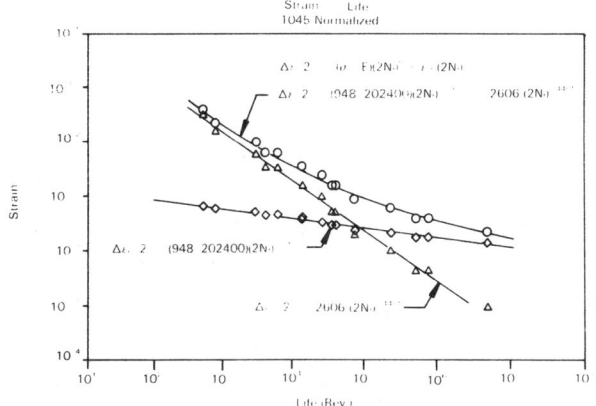

FIGURE 11. MATERIAL PROPERTIES AFTER ANALYSIS

```
DATA SUMMARY

MATERIAL FILE NAME    S1045A133
DATE ANALYSIS DONE    WED, JUN 29 1983

MODULUS                      E    0.2024E+06
YIELD STRENGTH @2% MPa            0.3815E+03
ULTIMATE STRENGTH, MPa            0.0000E+00
STRENGTH COEFFICIENT,        K    0.1185E+04
STRAIN HARDENING EXPON.,     n    0.2300E+00
TRUE FRACTURE STRENGTH            0.9845E+03
TRUE FRACTURE DUCTILITY           0.7100E+00
MEAN CYCLIC STR. COEF.,     K'    0.1253E+04
MEAN STRAIN HARD. EXP.,     n'    0.2075E+00
MEAN FATIGUE STR. COEF.    sf'    0.9481E+03
MEAN FATIGUE STR. EXP.,      b   -0.9235E-01
FATIGUE DUCT COEF.,        ef'    0.2606E+00
MEAN FATIG. DUCT. EXP.,      c   -0.4451E+00
```

FIGURE 12. DATA SUMMARY — 24 MAY 1983

SINGLE FILES

MATERIAL	Sigf'	b	ef'	c	K'	n'
1020 HR A	589	.0877	.2008	-.4434	817	.1996
1020 HR B	663	.0975	.4545	-.5087	772	.1916
1020 HR C	832	.1072	.2783	-.4643	1117	.2309
1020 HR D	581	.0854	.2125	-.4489	780	.1901
1025 HR A	935	.1017	.4772	-.4974	1087	.2046
1025 HR B	948	.0924	.3281	-.4946	1167	.1867
1035 CD A	837	.0735	.3966	-.5222	988	.1480
1035 HR B	897	.1113	.2900	-.4584	1211	.2428
1045 Norm A	948	.0923	.2606	-.4451	1253	.2075
1045 Norm B	1085	.1043	.3800	-.4913	1332	.2122
1045 CD F C	918	.0706	.2070	-.4516	1174	.1563
1045 HR D	1200	.1179	.4897	-.5324	1406	.2215
1080 HR A	1497	.1142	.6374	-.5709	2126	.2512

JOINING THE DATA NETWORK

Currently there are over 60 known specialized, computerized data bases in the United States and Western Europe. A significant number of these are presumably suitable for inclusion into the proposed Data Network. The concept of a distributed data network is to permit data base developers to maintain complete control over their information. They would, of course, have to permit access to their computers. As their data base is used, they would be compensated appropriately.

Any organization, corporation, technical society, trade association, government agency, research institute or university which desires to belong and complies with data quality standards, to be established, may be included in the network. Although the data network is being designed to handle numerical property data, other types of information such as loads, design algorithms, performance data, etc., could be included in the network if a market is identified.

RESPONSIBILITIES

This Materials Property Data Network is planned to be a cooperative effort by a number of organizations, each with a set of responsibilities as discussed below.

METAL PROPERTIES COUNCIL

The MPC, as motivator of the Materials Property Data Network, is establishing the management of the organization as directed by the Network's Board of Governors. The MPC, in accord with its role as a developer and evaluator of data, will provide appropriate committees which will coordinate activities in the industries covered by the network. MPC data bases and evaluations of data will be incorporated, and MPC will continue to identify needs for data.

TECHNICAL SOCIETIES

The cooperation and active participation of technical societies are essential for this development. Activities the societies can be engaged in are:

- Supply data bases
- Support committees for standardization of:
  - Material designation
  - Nomenclature
  - Material characterization
  - Test procedures
  - Data reporting
- Support committees for data handling
  - Data evaluation
  - Data analysis procedures
  - Data output
- Identify data needs
- Supply procedures to data usage

Many technical societies are already engaged in one or more of these activities and could devote their expertise to this undertaking.

NETWORK MAINTAINER

The network maintainer, yet to be selected, must provide the master control computer; the software for making the networking function, software for selection, analyses and output of data; maintenance to keep the network functioning properly; and a record maintained of the use of the network for proper billing of the users and compensating data base suppliers.

USERS

Users of the system must learn its capabilities and how to make optimum use of the system. They also have a responsibility to let their unsatisfied needs be known so these can be included as the system grows.

GOVERNMENT DEPARTMENTS AND AGENCIES

Several government departments and agencies have a need for material property information to carry out their mission. Therefore, the government has an interest in seeing that the system described here is successful and should support it with funds, data bases and technical expertise to assure that this system satisfies the needs of the many groups using material properties.

CURRENT STATUS

Various activities are under way as follows.

ORGANIZATION

The Board of Governors has been formed and is establishing policy and guidelines for the operation of the Data Network.

FUNDING

The Board of Governors is raising funds from industry to provide seed money for the project.

WORKSHOPS

Three application area workshops have been planned with the aid of the NBS, DOE, SAE, ASTM, ASME, ASM, AWS, AIAA, and EPRI. These workshops are to obtain user input on data needs including materials, properties, data standards, evaluation, analytical procedures, and data output. The workshops scheduled are for the following industries.

- Ground vehicles
- Aerospace
- Electric Power generation

Other workshops will follow.

COMMITTEES

Committees are being established to prepare specifications for the network and software.

- Material Designation, Nomenclature, Definitions
- Testing and Reporting Data
- Evaluation and Data Analysis
- Data Needs and Data Usage

Committees are being established to prepare specifications for the network and software.

CLOSING

Considerable activity is under way to create the National Materials Property Data Network, and it looks like we can achieve our first goal.

LITERATURE CITED

1. Panel on Mechanical Properties Data for Metals and Alloys, Numerical Data Advisory Board of the National Research Council, *Mechanical Properties Data for Metals and Alloys. Status of Data Reporting, Collecting, Appraising, and Disseminating,* National Academy Press, 1980.

2. Westbrook, J., *Feasibility Study of a National Materials Property Data Base,* Metals Properties Council, New York, New York, 1982.

3. Westbrook, J. H. and J. R. Rumble, Editors, *Computerized Materials Data Systems.* Proceedings of a Workshop devoted to discussion of problems confronting their development. Held at Fairfield Glade, Tennessee, November 7-11, 1982. Copyright 1983 by Steering Committee of the Workshop, NBS Office of Standard Reference Data.

4. Report by the Panel on Computerized Materials Properties Data of the National Materials Advisory Board of the National Research Council, National Academy Press, 1983 (In Press).

REVIEW OF EXISTING MATERIAL PROPERTIES COMPILATIONS

J.H. Westbrook ■ Materials Information Services, General Electric Research and Development Center, Schenectady, NY 12305

Material property compilations can be categorized into five broad groups: (1) handbooks, (2) printed graphical works, (3) data centers and projects, (4) data journals, (5) machine readable databases. Examples of each are given with special emphasis on machine-readable databases of engineering material properties. Fifty-eight have been identified in a recent survey. Current trends in material data compilations are reviewed with attention to gaps and other areas needing work.

INTRODUCTION

Several recent guides have been published devoted specifically to information sources on materials in addition to the inclusion of materials related sources in more general scientific and technical reference aids. Each has its own scope: Marvin and Sherwood (1) cover referral centers and publications as well as numeric data sources; Westbrook and Desai (2) emphasize numeric and graphical data; Behrens and Ebel (3) include significant material coverage in their multi-volume Physik Daten, segregated in the subsections - condensed matter, chemistry and materials; and Hampel et al (4), consolidating several earlier compilations, list over 120 machine-readable databases of materials properties. The several guides cited also include reference to other guides of both broader (e.g. all science and technology) and narrower (e.g. corrosion of metals or textile engineering and science) scope.

The present context does not permit re-listing of all of these information aids or even the subset for properties data alone. Rather, we will attempt here to categorize the various types of materials compilations which exist, give some illustrative examples in each category, describe some of their characteristics, and review the available auxiliary information services which facilitate their use. Reference may also be made to a general review of procedures and resources for the extraction and compilation of numerical and factual data in any technical field (5).

CATEGORIES OF DATA SOURCES

Five broad categories of data sources may be distinguished:

Handbooks	Printed compilations of numeric data usually in tabular form and often accompanied with tutorial text and graphical or pictorial illustrations
Graphical and Pictorial Reference Works	Printed compilations of non-numeric data usually in a standardized format, e.g. phase diagrams, transformation or TTT curves, Pourbaix charts, etc. for ready comparison
Data Centers and Projects	Activities of defined and limited technical scope, staffed by specialists in the area who are able to respond to phone or letter inquiries for specific data
Data Journals and Depositories	A relatively new medium especially adapted to data-archiving in print, tape or microform

Machine-readable Databases — Digitized information, usually stored magnetically on tape or disc, accessed either directly (on-line) or indirectly (off-line) with specially designed search and retrieval software

We will next examine each of these groups in more detail.

Handbooks {169}*

Examples

Thermophysical Properties of Matter	Touloukian and Ho**
Metals Handbook	ASM
Crystal Data, Determinative Tables	Donnay and Ondik
Aerospace Structural Metals Handbook	Belfour-Stulen Inc.
Alloy Digest	Alloy Digest, Inc.
International Plastics Selector	Cordura Publications
Optical Properties of Minerals: A Determinative Table	Winchell
Materials Safety Data Sheets	Nielsen and Kania (6) - GE
Dechema-Werkstoffe-Tabelle	Rabald et al (7) Dechema

Comments

Printed handbooks may focus on a particular class of materials, an application area, or a certain property or closely related family of properties. They offer the advantages of compactness of presentation, opportunity for browsing, and ready portability. On the other hand numerous disadvantages are also apparent. They readily become outdated and new editions are slow to appear and costly. Much valuable data never find their way into an appropriate handbook and remain scattered. Relationships between materials and property comparisons are difficult to deduce unless the editor has already chosen to highlight them. Locating individual data values depends on the accuracy and detail provided by the book's index. Searching for a material possessing a certain property or combination of properties can be an extremely tedious if not an impossible task.

Graphical and Pictorial Reference Works {37}

Examples

Phase Diagrams for Ceramists	Amer. Ceramic Society
The Particle Atlas	Ann Arbor Sci. Publishers
Atlas of Electrochemical Equilibria (Pourbaix Diagrams)	Pergamon
Atlas of Thermoanalytical Curves	Heydon & Sons
Casting Defects Handbook	Amer. Foundrymen's Soc.
Atlas of Isothermal Transformation and Cooling Transformation Diagrams	Amer. Society for Metals
Atlas of Crystal Stereograms	Earth Science Publ. Co.
Deformation Mechanism Maps - A Compilation	Frost and Ashley (8) Pergamon Press
Atlas of Hot Working Properties of Non-Ferrous Metals	Deutsche Gesellschaft für Metallkunde (9)

Comments

Graphical and pictorial reference compilations offer the advantages of substantial compression of the raw data and of permitting ready conceptualization of materials behavior and intercomparison of materials. Unfortunately, being bound books (sometimes looseleaf), they suffer the same disadvantages as

---

*The number shown here for this and parallel entries indicates the number of individual entries listed in 1977 by Westbrook and Desai (2). It is estimated that a more complete compilation today would reveal counts at least twice as high in each category. Nonetheless, the numbers cited give some idea of the magnitude and relative distribution of data sources.

**Full bibliographic citations will be found in Ref. (2) unless another reference number is shown.

printed handbooks. Digitization for computer storage and analysis is not so readily accomplished as with alphanumerics. Lastly, it is obvious that a particular mode of graphic representation must first be introduced that is so effective that it becomes a de facto standard and hence much data become available in that mode, or can readily be converted to it, before a significant compilation is feasible.

Data Centers and Projects {45}
Examples

Crystal Data Center	NBS
Molten Salts Data Center	Rensselaer Polytechnic Inst.
High Pressure Data Center	Brigham Young Univ.
Thermophysical Properties Data Center (CINDAS)	Purdue Univ.
Rare Earth Information Center	Iowa State Univ.
Plastics Technology Evaluation Center	Picatinny Arsenal
Machinability Data Center	Metcut Research Associates
Toxicology Information Program	National Library of Medicine
Metals and Ceramics Information Center	Battelle Memorial Institute

Comments

The primary goal of such centers or projects is the collection and critical evaluation of numerical data within their defined field, leading to the selection of "best" values within associated and defined limits of uncertainty. They are expected to be responsive to ad hoc inquiries and also to disseminate their work though published compilations, bibliographies and critical reviews. Such endeavors are no mere librarian or archival activity but require scientific knowledge and skill of the highest order. Close association between data evaluation centers and corresponding experimental research helps assure the critical judgment and insight essential to both. Indirectly, by communicating the experience from data evaluation, these centers can advance the level of experimental techniques, identify data gaps and dischordant values, and improve the reliability of new measurements. The National Bureau of Standards, through its Office of Standard Reference Data, has been a major influence in establishing, funding, monitoring and publicizing data centers and projects. Still other data centers are set up by trade associations, universities or private companies. From the foregoing remarks it may be realized that, to be effective, a data center must have true scientific expertise, be of at least a critical size, have immediate access to good laboratory facilities and experience continuity of operation. Unfortunately the last factor has often not been realized; more data centers are probably now defunct than are now in operation because of inadequate financial support. Ironically, this is not because of their high cost but because of poor perception of their true cost/benefits ratio.

Data Journals and Depositories {8}*
Examples
J. Physical and Chemical Reference Data
J. Chemical and Engineering Data
Bulletin of Alloy Phase Diagrams
Intl. J. of Chemical Kinetics
Organic Magnetic Resonance

Microfilm Depository Services (ACS)
Physics Auxiliary Publication Service (AIP)
National Auxiliary Publication Service (ASIS)
Depository of Unpublished Data (NRC-Canada)

Comments

This type of source is not intended to compete with the primary research or review journals but to afford a medium for dissemination and/or archiving of *data*. Some, such as J Phys. Chem. Reference Data, emphasize critically evaluated data and the methodology of evaluation. Others, such as the depositories, serve a purely archival function for raw data too voluminous for normal publication. Still others e.g. Calphad J., concentrate on methods for calculating data and comparing the results with experiment. In fields other than materials, such as the space, atmospheric or geosciences, magnetic tape storage of truly enormous volumes of data has become standard practice. Unfortunately, despite the prodigious recent gains in memory size, data compression, and storage cost reduction, there is no way all such data can be permanently archived and made accessible. In all cases of data journals and depositories, inadequate indexing prevents full utilization of these valuable resources.

---
*A more complete listing may be found in (5).

Machine-Readable Databases {28}*
  Examples (of chemical engineering interest)

Chemtran	Chemshare Corp.
Materials Property Package for Chemical Applications	U. Western Ontario
Physical Properties Data Services	Institution of Chemical Engineers
DETHERM-SDR	Dechema
"Thermo"	NBS-OSRD
RTECS (Registry of Toxic Effects of Chemical Substances)	NIH-NIOSH
EPIC (Estimate of Properties for Industrial Chemistry)	U. of Liege
IRSS (Infrared Search System)	NIH-EPA-NBS
TAP (Thermodynamic and Physical Properties Package)	U. Houston

Comments

Three studies - one by the Metal Properties Council, (11) an ad hoc Materials Data Workshop (12) and a committee report by the National Materials Advisory Board (13) - all show the great need for computer access to engineering properties information on materials. All find such a system to be technically within the capabilities of modern computerized information technology. The problem is that few of the several dozen materials properties databases built thus far are publicly available, and all are small and narrowly focused in subject coverage. Thus, while the applicability of the computer to materials information is well proven, an effective operating system of broad engineering utility is still lacking. To link the existing databases together, fill the data gaps present and supply needed supporting services will require a massive and continuing cooperative effort by a diverse group of stakeholders in such an information system. A report (14) has recently been prepared summarizing recent progress toward this end. Graham (15) has described the role and activities of the Metal Properties Council. Because of the importance of machine-readable databases and the current interest in the topic, their subject distribution, discussion of their capabilities and problems, and a projection of future trends of activity are presented in some detail in a latter section of this report.

AUXILIARY INFORMATION SERVICES

Data sources by themselves would be relatively ineffective were it not for the availability of a variety of auxiliary information services which enable us to be aware of, locate, identify and manipulate the data. We will discuss here and give examples of these types of services.

Structure and Nomenclature

A number of resources exist, both in print and on-line, intended to assist the seeker of data by providing identification keys and cross-references from common and trade names, industrial numbering systems and "universal" numbering systems such as the Chemical Abstracts Registry Numbers or the ASTM/SAE Universal Numbering System. In addition, in the chemical field several different systems have been devised to permit on-line searching by chemical structure or sub-structural fragment as reviewed by Rush (16). Examples of some of the structure and nomenclature auxiliary services are listed below. Further details of the operation and application of certain of these systems may be found in the references cited in the table.

Structure and Nomenclature Resources

CAS ONLINE	Chemical Abstracts
SANSS (17)	NIH-EPA
ANSA	Inst. for Scientific Info.
DARC (18)	Centre Info. Doc. Automatique
SHE	Engineering Info., Inc.
Metals Datafile	ASM
IMMCR (19)	GE
Engineering Alloys (20)	Van Nostrand-Reinhold
Metallic Materials Specifications (21)	Spon

*This category has been growing very rapidly of late. As previously noted, Hampel et al (4) now list more than 120 machine-readable materials properties databases.

Data Guides and Indices

Both retrospective and coincident indexing of data are essential to the problem of determining whether property x of substance y is known and to what degree of reliability, since conventional bibliographic abstracts, either print or on-line, are quite inadequate to determine with certainty whether quantitative numeric data are included in the original source. In contrast to data abstracts, journals and services discussed next, the guides listed immediately below direct the reader only to other data compilations and not to the original sources. They are very valuable, all the same.

Data Guides and Indices*

Critical Surveys of Data Sources	NBS 396-1,2,3,4
Data Sources for Materials Scientists and Engineers	Westbrook and Desai
A Brief Guide to Sources of Metals Information	Hyslop
Thermophysical Properties Research Literature Retrieval Guide	Touloukian, et al
A Guide to Sources of Information on Materials	Marvin and Sherwood
Permutted Materials Index - Alloy Data Center	NBS 324
Directory of Data Sources for Science and Technology	CODATA
Inventory of Data Sources in Science in Technology	UNESCO/ CODATA
Environmental Data Index	NOAA
Chemical Information Systems	Ash and Hyde (22)
Compilations of Alloy Phase Diagrams and Related Data	Kahan, et al (23)

Data Abstracts

Another class of service is being provided by those continuing publication series which partake of aspects of both the bibliographic abstract journals and the pure data collections. These reference works extract newly reported data from primary sources and republish them with just such descriptive information as is required for minimal interpretation. The data are frequently brought to a standard format but are only rarely evaluated. Examples are listed in the table below. Metals Datafile is somewhat different than the other entries shown. This resource is the product of a recent collaboration by the American Society for Metals and SDC, Inc., to add the numeric equivalent to ASM's Metadex bibliographic abstract file on metals and alloys. This new service is in part retrospective (covering leading handbooks, reference works and data collections in the metals field) and in part coincident (adding specific data to currently produced abstracts). The on-line aspects and the expanded search capabilities are attractive, but Metals Datafile has not yet achieved a broad and enthusiastic response from the materials community because of its use of unevaluated data, the arbitrariness of the selection of data for inclusion, and the incompleteness of its coverage. Improvements can undoubtedly be expected in the future.

Data Abstracts**

Metals Datafile
Mechanical and Corrosion Properties
  Key Engineering Materials
  Single Crystal Properties
Diffusion and Defect Data
Structure Reports
Mössbauer Effect Data Index

MACHINE-READABLE MATERIALS DATABASES AN IN-DEPTH EXAMINATION

The most comprehensive directory of machine-readable materials properties databases is that of Hampel et al. (4) Some insight as to the nature of the coverage is provided by the analysis shown in Table I which follows a breakdown originally suggested by Hilsenrath (24). The 58 databases in the last category of engineering properties may be further broken down as shown in Table II. It must be emphasized that in contrast to the other entries in Table I, these engineering databases, while machine-readable, are rarely available to the public on-line. Some other observations are also notable. Foreign developers of such databases have been as active as

---

*Except where otherwise noted full bibliographic citation of these sources appears in Ref. (2) or (5).

**Except for Metals Datafile which has not yet been fully described in the literature, bibliographic citations for the other examples shown are included in Ref. (2) and (5).

Table I

Machine-Readable Databases for Materials Properties
Included in the Directory of Hampel et al (4)

Systems for Identification of Unknown Substances	19
Systems for Properties of Pure Substances and Mixtures	20
Systems for Metallurgical Calculations	5
Systems for Thermodynamic and Thermochemical Properties of Individual Substances	6
Systems for the Properties of Plastics	2
Systems for Chemical Process Simulation and Design	11
Machine-Readable Files of Engineering Data on Materials	58
	158

Table II

Machine-Readable Databases on Engineering Properties
of Materials Identified in Metal Properties Council Survey of 1982 (11)

By Country	By Material	By Property
30 US		
7 U.K.		
4 Japan		
3 France		
3 Canada		
2 W. Germany		
2 Poland		
1 E. Germany	34 Metals	40 Mechanical
1 Netherlands	4 Composites	7 Process-related
1 Italy	2 Semiconductors and Electronics	2 Thermal
1 Austria	3 Processing	2 Electrical & Electronic
1 Yugoslavia	8 Polymers	
1 Czechoslovakia	1 Fibers	5 Other
58	58	58

the U.S. and a heavy topical concentrations on the mechanical properties of metals is also evident. Of the 63 databases in the first six categories in Table I, 37 appear in the latest edition of the Cuadra directory (25); these latter are broken down in another way in Table III. These several analyses illustrate the diverse origins and motivations for the creation of machine-readable files as well as the unevenness of their present coverage of the field.

Turning now to the capabilities of machine-readable files, it is readily apparent that the excitement and interest attending their development arises not from the simple volume compression of data and ease of access

Table III

Textual and Numeric Databases
Included in Cuadra Directory
of On-Line Databases (1983)

Spectra	7
Crystallographic	4
Properties	19
Toxicological	4
Thermochemical and Thermophysical	10
Nuclear	1
Electronic	1
Physical and Mechanical	3
Structure and Nomenclature	7
	37

they afford compared to conventional compilations, but rather the new capabilities available which cannot be matched by printed sources. Some of these are summarized in Table IV. A more extensive discussion of the meaning and application of these capabilities appears in Ref. (11), (12) and (14).

Table IV

Capabilities of Computerized
Materials Data Systems

o File inversion with -
    relational constraints
    property weighting
    sensitivity analysis
o graphical display and comparison
o rank ordered files
o derivative properties
o automatic unit conversion
o statistical analysis
o anomaly identification
o interpolation, extrapolation and other
    estimation techniques
o direct link to CAD/CAM
o responsive system development
o facile expansion and updating
o charges proportional to use
o direct channel for user feedback

Similarly, Table V summarizes the problems associated with existing machine-readable databases on materials, either singly or collectively. It is apparent that in the main these are not technical issues but rather relate largely to the content, reliability and portability of the data themselves and the awareness and appreciation by the potential user of the existence and value of such systems. Again, these matters are more fully elaborated in Ref. (12) (13) and (14).

Table V

Problems with Existing
Machine-Readable Databases

o gaps in materials/properties coverage
o few evaluated data
o lack of standardization
    (materials designations, test methods,
    properties)
o little net-working capability
o referral capabilities undeveloped
o machanics/economics of conversion
    of print to machine-readable form
o lack of user awareness
o unperceived value of reliable information

SOME CURRENT TRENDS
IN MATERIALS DATA PROGRAMS

Several trends are apparent in current materials data activities which warrant comment.

"Gateway" or One-Stop Information Shopping
Concept

This concept, in which a single phone call would put the user in touch with a wide variety of information resources, first arose from a 1980 NDAB study (26) and has since been endorsed and expanded upon in several different studies (11-13). In addition to vastly improved user convenience and the opportunity for overcoming the persistent problem of subcritical database size and scope, the "gateway" concept also affords benefits of standard support programs, on-line directories to other data sources and built-in tutorials. Still other benefits include:

- Focused expertise in building and maintaining individual databases
- Easier access and application by users
- Easier homogenization of database structures
- Facile expansion
- Ready tailoring to serve different user markets
- Easy exploitation of existing database construction
- Lower total cost
- Decreased redundancy of effort

Increased Emphasis on Evaluation Programs

It is being realized more and more that a primary reason for the low level of on-line use achieved by technical numeric data systems is the low, or at least undefined quality level of the data included. Data evaluation is a costly process and requires the attention of true experts in the particular materials/property field in question. Materials groups active in this area are growing and now include the Metal Properties Council, the Electric Power Research Institute, AIChE through its DIPPR program, CINDAS at Purdue University and CODATA. The last group have been interested in refining and publicizing technique for data evaluation (27) as well as in performing the evaluation work itself in a wide variety of disciplinary areas. Of equal importance to data evaluation is full characterization of the data once it is to be entered in the information system. For each number there should be shown a) the type of data (measured, evaluated, predicted, standard, etc., b) the estimated accuracy and c) the ultimate source of the data and its evaluating or certifying authority.

Improved Directories of Data Sources

The most complete directory now available of machine-readable materials data compilations is that of Hampel, et al (4) previously cited. Desirable next steps are to make such directories available on-line for easy access and updating and to provide more detail as to the operational features, languages, access protocols, explicit materials/properties coverage, etc. Such efforts are just now getting underway.

"Super-Index"

This term refers to a project of CRC Press to merge the indices of both its own several hundred technical handbooks and monographs and those of other technical publishers into a "Super-Index." The first version of such a file has been built and is now available on-line. In future years its utility should be improved by increasing comprehensiveness, decreased ambiguity in indexing terms and additional intellectual effort in structuring and cross-referencing within the "Super-Index."

Cooperative Activities

Data compilation, evaluation and computerization are such large, complex and costly activities that new modes and higher levels of cooperation than ever before will have to be effected: between data generators and data users, between the public and private sectors, between technical disciplines and between the nations of the world. Such cooperation would have many benefits to the participating parties including: cost sharing; focused expertise; decreased redundancy of effort; complementarity of needs, skills and resources; ability to marshall a critical level of effort; opportunity for standardization and compatibility; and enhanced promotion and training. This subject has been elaborated further in two recent reports (12), (28).

"Expert" Systems

It is widely held that the next generation of computerized information systems will have a new level of capability beyond simple search, retrieval, manipulation and display of data. That is, a combination of stored logical programs, "rules", and the result of an interactive "dialogue" between the computer and the user will yield near-human-like performance in solving problems such as materials selection or failure analysis. Duda and Shortliffe (29) have summarized recent progress in this field and describe applications in medicine, geology and mass spectroscopy; the first attempt to do something similar in the materials field is the work of Weiss, Robinson and Sibert (30) of Syracuse U. using the LOGLISP computer language.

Computer-Aided Engineering (CAE)

Computer-aided Design and Manufacturing (CAD/CAM), which are becoming so prevalent today, are still largely restricted to automation of drafting, geometric conceptualization and manipulative functions. Even when the computer is employed in a calculational mode as with design algorithms, the basic data are almost always input by hand from some external printed store. All this will change as more materials properties data become available in machine-readable form and a complete integration of all engineering functions is made possible. With this will evolve an electronic work station for engineers which will permit not only information access and display but also calculation, "drawing" generation, performance prediction etc.

Logical Structuring of Knowledge

The burgeoning growth of computerized information systems of all kinds, not just "expert" systems, is forcing increased attention to the formal logic and structuring of knowledge in every disciplinary field. Early efforts along this line in the materials field have been made by Allen and associates at Brigham Young University. (31)

LITERATURE CITED

(1) Marvin R.S. and Sherwood, G.B., "A Guide to Sources of Information on Materials" Handbook of Materials Science, CRC Press, Cleveland, OH, v3, p603-627.

(2) Westbrook, J.H. and Desai, J.D., "Data Sources for Materials Scientists and Engineers," Ann. Review of Materials Science 8 (1978) 359-422.

(3) Behrens, H. and Ebel, G., "Physik Daten" Zentralstelle fur Atomenergie - Dokumentation, v3-1 (1976), v3-2 (1977), v3-3 (1978), v3-4 (1979).

(4) Hampel, V.E., Hilsenrath, J., Westbrook, J.H., Gaynor, C.A., and Johnson, P.S., "Directory of Databases for Materials Properties" August 1983, Lawrence Livermore Lab. Report UCAR 10099.

(5) Westbrook, J.H., "Extraction and Compilation of Numerical and Factual Data" AGARD Lecture Series #130, "Development and Use of Numerical and Factual Data Bases," Gaithersburg, London and Lisbon, Oct. 1983.

(6) Nielsen, J.M. and Kania, C.J., "Material Safety Data Sheet Collection" General Electric Company (1983).

(7) Rabald, E., "Dechema-Werkstoffe-Tabelle," Deutsche Gesellschaft für Chemisches Apparatewesen, Frankfurt am Main (1973).

(8) Frost, H.J. and Ashby, M.F., "Deformation-Mechanism Maps" Pergamon Press (1982).

(9) _____, "Atlas of Hot Working Properties of Non-Ferrous Metals" v1 and 2, Deutsche Gesellschaft für Metallkunde (1979).

(10) Fivozinsky, S.P., ed., "Technical Activities 1982, Office of Standard Reference Data" NBS-IR-83-2661 (1983).

(11) Westbrook, J.H., "Feasibility Study of An Inter-Society Computer-based Material Property System." Report to the Metal Properties Council, Dec 1982, 91 pp.

(12) Westbrook, J.H. and Rumble, J.R., "Computerized Materials Data Systems" Proceedings of the Materials Data Workshop, Fairfield Glade, TN (1982) 133 pp.

(13) Brown, W.F., et al, "Materials Property Data Management - Approaches to a Critical National Need," NMAB Report-405 (1982).

(14) Westbrook, J.H., "Progress Toward a Coordinated System of Databases Covering the Engineering Properties of Materials" AGARD Lecture Series #130 "Development and Use of Numerical and Factual Data Bases" Gaithersburg, London and Lisbon, Oct 1983.

(15) Graham, J.A., "The Metal Properties Council's Activities on a National Materials Property Data Network" AIChE meeting, Denver, Colorado, Aug 1983.

(16) Rush, J.E., "Handling Chemical Structure Information" Annual Review of Info. Sci. and Technology 13 (1978) 209.

(17) Heller, S.R., Milne G.W.A., Fein, A.E., Frees, E.F., Marquart, R.G. McGill, J.A. Miller, J.A. and Spiers, D.S., "NIH-EPA Structure and Nomenclature Search System" J. Chem Info. Computer Sci. 18 (1978) 181-186.

(18) DuBois, J.E., "French National Policy for Chemical Information and the DARC System as a Tool for this Policy" J Chem Doc 13 (1973) 8-14.

(19) Arcuri, J.V. and Potts, D.L., "International Metallic Materials Cross Reference" 2nd ed. 1983, General Electric Co., Schenectady, NY.

(20) Woldman, N.E. and Gibbons, R.C. Engineering Alloys, New York: Van Nostrand-Reinhold. 1427 pp. 5th ed. (1973).

(21) Ross, R.B., Metallic Materials-Specification Handbook. London: E&F Spon Ltd. 833pp. (1972).

(22) Ash, J.E. and Hyde, E. eds, Chemical Information Systems, Ellis Harwood Ltd. Chichester, England and Halsted Press, New York 309pp (1975).

(23) Kahan, D.J., Harris, J.F. and Bennett, L.H., "Compilations of Alloy Phase Diagrams and Related Data," Bull. Alloy Phase Diagrams 3 (1983) 417.

(24) Hilsenrath, J., "Summary of On-Line or Interactive Physico-Chemical Numerical Data Systems," NBS Technical Note TN1122, Oct 1980.

(25) _____, "Directory of On-Line Databases" 4 #3 Spring 1983. Cuadra Associates, Inc., Santa Monica, CA.

(26) Graham, J.A., et al, "Mechanical Properties Data for Metals and Alloys - Status of Data Reporting, Collecting, Appraising and Dissemination," NDAB-NRC (1980) 24 pp.

(27) Rossmassler, S.A. and Watson, D.G., Data Handling for Science and Technology - An Overview and Source Book, North Holland Publ. Co., Amsterdam, 184 pp (1980).

(28) Westbrook, J.H., "Cooperation in Developing Computerized Material Databases" Data for Science and Technology, Proceedings of 8th International CODATA Conf., Oct 1982, Jachranka, Poland, North Holland Publ. Co (1983) 91-98.

(29) Duda, R.O. and Shortliffe, E.H., "Expert Systems Research" Science 220 (1983) 261.

(30) Weiss, V., private communication.

(31) Allen, D.K. and Smith P.R., "Engineering Materials Taxonomy" Monograph #4, Computer Aided Manufacturing Laboratory, Brigham Young U. (1982).

DATA FOR ENGINEERING DESIGN

J.A. Shaw ■ ESDU International Ltd., 251/9 Regent St., London W1R 7AD, England

---

The paper discusses the effect that large mainframe programs have had on the way engineers view empirical data, and suggests that the inertia inevitable in the process of updating large programs means that engineers now use out-of-date data for longer than in the past. The need for physical properties data is discussed and comments are made on the accuracy claimed for experimental measurements by authors. The requirement to obtain accurate data, or to know the tolerance on values used, is established and the method by which ESDU arrives at each value for physical properties data is explained.

The Need

As an organisation dedicated to the provision of engineering data, we spend a great deal of our time discussing with practising engineers the question of what data are required in design. In recent years we have found more and more that there is a lack of appreciation that data are a necessary part of any design process. Probably the single most important contribution to that situation has been the advent of the computer-technology world. Much of today's design is done using large mainframe programs of great complexity. The engineering designer has developed into a keyboard operator and many of those to whom we talk seem to have little if any understanding of the basis of the code that is providing the results they get on their print-outs. So it is not too surprising that they should not realise that many routines in those programs do use data of various kinds.

Most engineers realise they need some kinds of data; materials and physical properties are recognised by all of us as required information, probably because we know there are few possible ways of predicting a proof stress or a thermal conductivity. But there seems to be developing a misapprehension that the rest of the design information can be calculated by computer as required without further reference to continuing research, except perhaps when someone makes a particularly far-reaching theoretical break-through. In fact, of course, these programs contain many sub-routines based entirely on empirical correlations. For example, calculations of heat transfer coefficient or pressure loss, even in single-phase flow, are based entirely on empiricism. This lack of understanding of the basis of the day-to-day bread and butter of the designer leads to a most dangerous situation. Not only is, in fact, new information constantly generated by research programmes throughout the world that means that the correlations built into the software will gradually drift out-of-data and the products based on them will become uncompetitive (if any competitor up-dates his data, of course). But, moreover, within any Company there is a continuous pressure to extend the range of application of the products, a drive to make them bigger, faster, cheaper. The result of that pressure is relentlessly to push further and further the ranges of the parameters within the programs, to higher heat fluxes, bigger throughput, higher stresses (or bigger components). Eventually, and certainly quite unbeknown to the user of the software, who by now is someone quite remote from those who originally wrote it, some limitation on a parameter within that software will be exceeded. The consequences of that are unknown. They may, or may not, be serious. But one thing is certain and that is that no-one will know until it is too late.

A major contribution to the lack of understanding of the need for data is the finite-element package. A superb tool in the hands of the intelligent designer, it can snap at fools who have a blind faith in its output. But leaving aside the very real question of the black art involved in modelling the component or structure, there is another major problem in its use. The package cannot assist the designer in any way to understand trends and exchange rates except by a pure trial and error process. To attempt to design for least weight, or indeed any other optimum, with a finite element package alone and no fundamental stressing knowledge will not take anyone too far.

So in fact the computer has in no sense eliminated the need for sound engineering data, not even in strength analysis, let alone in heat transfer or in physical properties. It has only made it a great deal more difficult for us to ensure that those data are kept up-to-date because the designer is no longer aware that they are there. His reliance on the big program means that it is that that has to be up-dated whereas previously, when he himself was aware of the correlations he used, he was more likely to recognise the need to update them himself. And up-dating a large program is a great deal more difficult than throwing away one graph or table and replacing it with another.

The Data

Before the large mainframe programs existed, the engineer made use of data in all his design calculations. Where did they come from? Generally there were a number of sources: college notes rank high, old college text books, and various handbooks like Machinery Handbook and Perry. Then, of course, as the designer developed in experience, he stopped looking for data. He used his own collection, built-up in the early days as he learnt his trade, and used thereafter as if engraved on a stone tablet. Such data were, and are, very good. But they cannot by definition by up-to-date. And so they cannot be adequate as a basis for safe and reliable design or indeed for competitive design in the economic environment in which we find ourselves today.

If the data on which you design are known by you to be sound, you can have the confidence to avoid over-design, which is expensive in materials, and you can avoid under-design, which leads to failures that can have expensive consequences in a chemical plant far exceeding the mere cost of replacing a failed component. Because you have confidence that your pressure loss, heat transfer, or stress calculation is correct, you can cut your safety factor down to that required to cover any remaining uncertainties.

So to summarise, good design data are still today as essential as they have always been to maintain the competitive edge in a commercial world. The inertia of the system is now greater than it has ever been but the need remains, and you must maintain a constant pressure and vigilance to ensure that the tools available to you are as good as they can possibly be. However, before introducing one solution to that problem, I would like to turn to the question of physical properties data.

Physical Properties Data

The correlations of heat transfer and pressure drop are usually developed in a form that requires a knowledge of the physical properties of the liquids or gases in order to use them. Thus, it is of little significance to have available to you a much improved correlation of two-phase flow pressure drop if those physical properties data are not of a comparable accuracy.

It must be recognised that taking data from a handbook is not going to give you the accuracy you need, generally speaking. So what is the alternative? You can go into the research literature to find out what data exist, probably using one of the large mainframe retrieval systems, indeed almost certainly using Chemabs. And you will obtain a very large number of references, in all probability, unless your particular need is of a very esoteric nature. Those references will all have, as guides, abstracts. Unfortunately, for the needs of the engineering designer far too little thought (and, perhaps more relevantly, engineering know how) goes into selecting even a title, let alone an abstract. Thus as a guide to determining the relevance of a particular paper, the abstract is rarely adequate. So, in all probability, in order to be sure you leave no stone unturned, you now need to collect together all those papers, read them all, digest them, and eventually arrive at an appropriate conclusion. Reaching that conclusion may well be difficult, moreover, because it is not unusual to find quite significant discrepancies between the various sources, as we shall see later. So after that effort, and obtaining all the papers can take an elapsed time of over a month without any great difficulty, our designer has now to spend time resolving that discrepancy; he has to investigate the experimental techniques to see whether one result is likely to be more reliable than another, he has to consider the possible effect of impurities, he may even need to go and consult the experts in the field. Is it any wonder that if the designer ever enters into such an exercise once, it is the last time he does it?

Next time he takes a guess, he uses his handbook, he takes manufacturers' guidance, he does something relatively quick, and he increases his safety factor. If he did not do that, his tender would always be a year or two too late.

But that was a relatively simple problem. In practice the chances are that the designer needs values at temperatures or pressures at which to date there exist no measurements in the literature he has located. Now what does he do? He can only take his usual route; he can have a guess and increase his safety factor yet again. But this business of guessing is risky. It requires experience of the problem, and physical properties are at best totally ancillary to, and at worst remote from, the designer's real interests in making a successful chemical plant. Thus ideally one would think that the person to make the guess would not be the designer but a researcher in the field of physical properties data. Unfortunately, they are very often the last people prepared to hazzard a guess. Not only do they know too much about the potential pitfalls, their training and background is usually scientific rather than engineering. Thus the process of making an informed extrapolation or interpolation is an anathema to them.

There is yet another problem. You may have located one report that just happens to contain exactly the data you were seeking. The researcher who wrote the report quotes an accuracy of +2 per cent. So, great, now you have the answer and you know the tolerance on it. You do not need any safety factor. But unfortunately that is not true. Researchers seem to be capable of looking at their own achievements through rose tinted spectacles. They have not surprisingly a predilection to their own results over those of other researchers. Not only, therefore, if they compare their results with others drawn from the literature, do they always draw the curve biased towards their own data (if not through them), but they will quote accuracies for their own results that those plots of available data quite clearly give the lie to.

Please understand that I am not making here any holier than thou accusation; what I am saying is, we all recognise, simply human nature. And it is in no sense an unsupported statement. I am used to plotting out data and seeing a scatter of perhaps ± 25 per cent, where all the data themselves are claimed to have individually accuracies of ± 2 or 3 per cent. Nor is it only a personal view. R C Dean, editor of Fluids Engineering for example, has gone on record (Ref.1) with the following statement:

> "My watch of authors uncertainty estimates streaming by for the last two years integrates to a general conclusion that the average uncertainty in output results of fluids experimentation is about ± 1 per cent. I know that is not true. (author's italics) I would guess that the average is more like ± 10 per cent, with uncertainty of ± 50 per cent pertaining to at least 20 per cent of the data."

And he goes on to quote R W Roberts, Director of the US National Bureau of Standards, as estimating that,

> "half or more of the numerical data published by scientists in their journal articles are unusable because there is no evidence that the researcher accurately measured what he thought he was measuring or no evidence that possible sources of error were eliminated or accounted for.".

There is yet another potential pitfall for the unwary. If you are a designer, and you have collected together all your data from the literature, and you have plotted them, and you have found a considerable scatter, there is a natural temptation to assume that the data remote from the mean must be in error and should simply be neglected. Or you may decide to approach the problem systematically, and simply to discount all the elderly data on the not unreasonable assumption that modern results using modern techniques are the more reliable. But look at Figure 1 taken from the Jamieson's paper (Ref.2). Age is clearly no criterion for discounting data, and indeed looking at that figure, who would not be baffled to make an informed guess without a considerable background of knowledge? It is perhaps of some interest here to have a look at the variation of thermal conductivity with temperature for some fluid normal alkanes obtained using a method of simply drawing the best mean line through sets of experimental data for each compound, albeit weighted according to an informed assessment of their experimental accuracy. I am sure we all expect a reasonably systematic trend of the data with molecular structure and therefore know instinctively that that cannot be right. But how to correct it is yet another problem. (Figure 2).

The ESDU Solution

The process of validating data that we have developed over the years has been continuously revised to take account of the problems I have outlined as we have painfully over that period recognised them and their importance.

Firstly a few words about ESDU. It is a unique organisation in that its sole raison d'etre is the provision of evaluated engineering data of all kinds, not only physical properties data. It has no research facilities and its total independence from research as a business ensures its unprejudiced approach to the information it evaluates. It has a full time professional engineering staff whose sole responsibility is the evaluation of data in which they therefore develop a wide experience and capability. They take a limited individual topic, be it thermal conductivity of alkanes or heat transfer inside tubes, they collect all the published information, assess it and distil it into a working tool for direct application by the working engineer.

But that on its own is not enough; an individual's assessment of the information can only be as good as his knowledge, experience and skill can make it. In an ideal world what is really required is for two or three people to go away and undertake that process of assessment and then for yet another to take those, undoubtedly, two or three different conclusions to obtain a final agreed consensus. Clearly that would not be a practical approach to produce anything in a reasonable timescale economically, but ESDU does the next best thing. We collect together a group of experts from industry, research and academia who meet together regularly with the staff man to discuss and assess the results of his work. They reach that consensus with him before the assessed data can be issued. This group of experts, who include corresponding members from abroad (and some attendees from Europe) bring an additional advantage. They often bring to ESDU data that are unpublished, and ESDU's uncompetitive position with the majority of other organisations enables it to obtain information that any other organisation attempting to do this work in association with other activities would not be able to get.

The Correlation of Physical Properties Data

In assessing physical properties data, what we have done has been to take an equation that has been found to represent well the variation of a property such as thermal conductivity with temperature. Over the years that equation has inevitably been refined, but the principles are not changed. That equation will contain one or more constants whose values are fixed for a particular compound and the process of correlation is basically one of determining appropriate values for these. The available data, having been individually assessed for each compound, are then fitted by the chosen correlating equation, and the behaviour of the values of the constants with molecular structure is examined. There are generally for most groups of organic compounds a number of ways in which those correlations can be studied. For example, with each property one starts with the straight chain alkanes. But, thereafter, that correlation provides a base so that one can develop rules for the substitution of hydrogen atoms by other atoms or groups, and thereby can arrive at the effect on the coefficients of molecular structure variation in two dimensions. For example, if one considers the ketones, then one can view the variation of the coefficients both with increasing carbon number and also from an established base provided by the alkanes by the substitution of an oxygen atom for a CH_3 group in the appropriate homomorph. The great advantage of this approach is that it makes available to the correlator the data for all the compounds and not just those for an individual compound. As a result, it not only enables extrapolation with temperature to be made with reasonable confidence, it allows prediction of values for compounds in a series for which no experimental results at all are available. It is interesting to note that when the technique was relatively unproved in the early days of the work, the Committee expressed some concern about the extent to which the staff had interpolated and extrapolated among data for thermal conductivity of the esters. They asked that some specific experiments be undertaken on three compounds strategically chosen at the extremes of the correlation to verify the predictions. The results obtained were all within ± 5 per cent of the predictions and in most cases actually came within ± 2 per cent.

The approach that has been adopted here is pragmatic. We have not sought as others have to develop a group contribution method in a pure scientific sense; we have merely assumed that nature is reasonably orderly and that each step is related to the previous ones in some logically progressive way without fixing the value of that step from any previously derived information. For engineering purposes, however, we can with confidence say that it works.

Conclusion

Engineers rarely compare the cost of using good data with that of bad or inferior data. They almost never record the time spend looking for data. The best data are deceptively simple but like a wrench - if it is not looked after it becomes damaged or rusty. Engineers service and maintain their calculators and computers but data lie forgotten until they are needed and no one recognises the dangers of that situation in terms of safety, reliability or profitability.

The ESDU "validated" data, as you have seen, overcome all these problems, by the way in which they are produced and updated and through the advisory service that is provided to support the data.

LITERATURE CITED

1. Dean, R.C., "Truth in Publication," J., Fluid Engineering, pp. 270, (June 1977).

2. Jamieson, D.T., "Thermal Conductivity of Liquids" J., Chemical Engineering Data Vol. 24, No. 3, pp. 244-246 (1979).

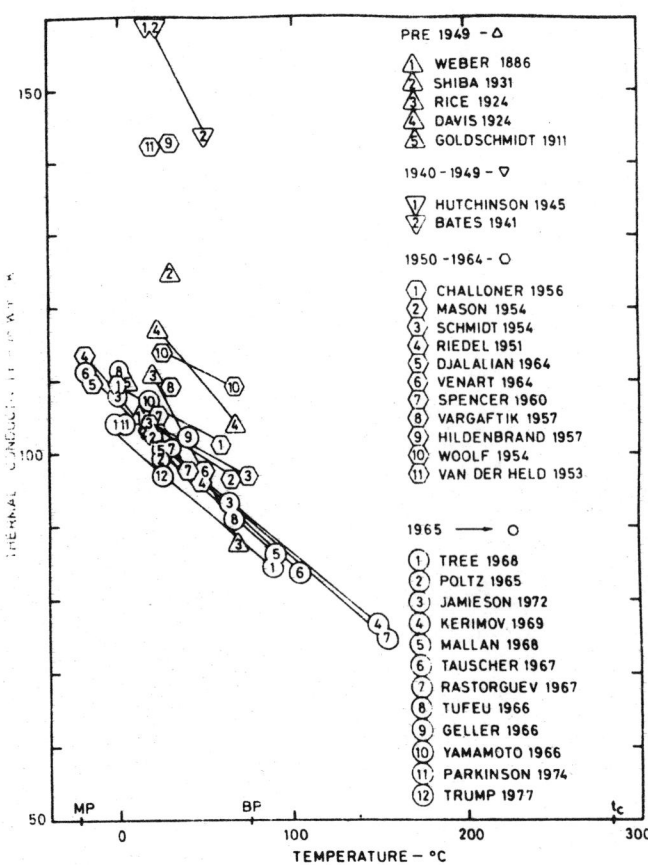

Figure 1. Carbon tetrachloride, thermal conductivity as a function of temperature.

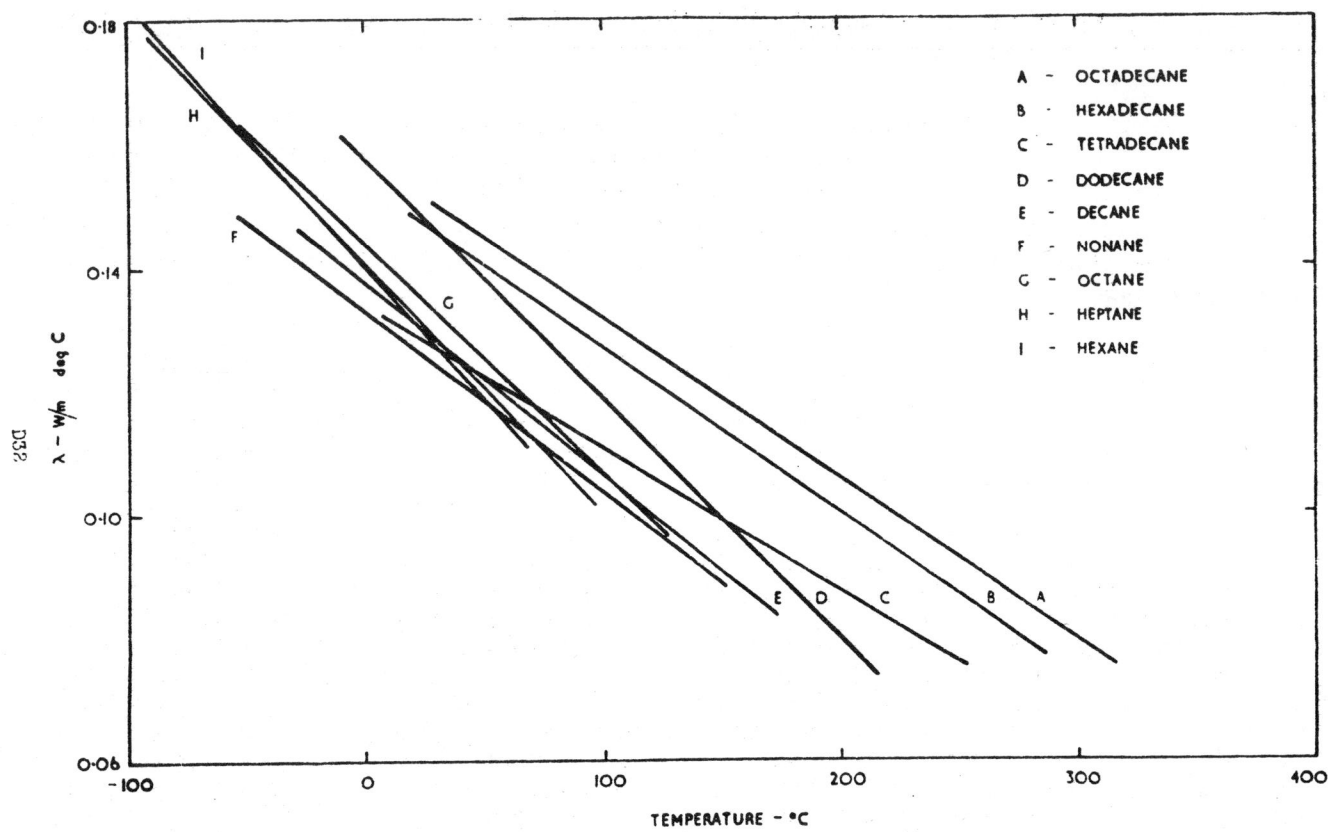

Figure 2. Thermal conductivity of liquid normal paraffins.

FACILITY FOR THE ANALYSIS OF CHEMICAL THERMODYNAMICS

W.T. Thompson ■ Royal Military College of Canada, Kingston, Ontario K7L 2W3
A.D. Pelton and C.W. Bale ■ Ecole Polytechnique de Montreal, Box 6079, Station A, Montreal, Quebec H3C 3A7

F*A*C*T is an on-line thermodynamic database computing system containing data on several thousand compounds and aqueous ions and a few hundred binary solutions. The bulk of these data have been derived from standard sources. It includes a suite of user-friendly programs which permits calculation of heat balances, heterogeneous equilibria, isothermal predominance and E-pH diagrams, and binary and ternary temperature/pressure-composition phase diagrams. The system is available by telephone throughout North America via TELENET or DATAPAC and may be operated by personnel untrained in computer programming.

INTRODUCTION

F*A*C*T, the Facility for the Analysis of Chemical Thermodynamics, is an interactive database computing system for performing classical thermodynamic calculations. It is a user-friendly system in which users with no prerequisite knowledge of computer programming connect their terminals by telephone to a host centre and then converse with the system. A broad range of calculations can be performed from the simple computation of extensive state property changes for a chemical reaction to the construction of a multicomponent phase diagram. All of the computed tables and diagrams are immediately printed on the user's terminal in reply to his continual thermodynamic inquiries. Data automatically drawn from the main database may be used in combination with a private user-managed database accessible only by that particular user. The intent of the system is to put engineers and scientists in closer contact with thermodynamics by replacing the tedium of traditional data retrieval and calculational procedures with the more creative aspects of thermodynamic analysis.

SYSTEM ARCHITECTURE

Figure 1 shows schematically the configuration of the system. Up to 140 terminals may concurrently share F*A*C*T software. A wide variety of terminal types is supported at 300 or 1200 BAUD. Telephone connection using an acoustic coupler and conventional telephone (or the equivalent) is made via TELENET in the United States or DATAPAC in Canada. With such systems, performance at the most remote site is virtually the same as that at a terminal located adjacent to the host computer. Time sharing management on the computer is accomplished by MUSIC. The system, however, is sufficiently portable that other combinations of mainframe and time sharing system may be substituted.

After entering F*A*C*T, the user is presented with a menu of programs in ready-to-run form. During the execution of a particular program, there are interruptions while the user is prompted to supply computational directives. The prompting is in English or French depending upon the preference when signing on. Illogical input results in error messages and the user is always given the opportunity to reenter the required input such as the correct formula for a compound. The numerical data files are searched when required in a completely automatic fashion requiring no intervention by the user. There is no batch mode; all of the output is immediately directed to the user's terminal. This conversational mode insures good coupling of human decision making and machine labour and reduces the expensive generation of voluminous unnecessary data often associated with batch mode.

Subscribers to this system are invoiced by the host centre in direct proportion to use and there is no monthly minimum. This includes telecommunication expenses which are made distance independent by accounting procedures at the host centre. Reduced rates apply to educational institutions. The following sections highlight some of the capabilities of F*A*C*T.

REACTION

Figure 2 is indicative of the type of problem that can be handled with the REACTION program (1,2) although it is a particularly simple example so that attention can be focussed on the program. The user enters a balanced equation in this case representing the combustion of carbon monoxide with the stoichiometric amount of oxygen for complete combustion to produce carbon dioxide. The notational scheme is much like conventional chemical notation except that stoichiometric integers or asterisks must be used to separate alphabetical combinations that represent elements. This avoids for example ambiguity between cobalt and carbon monoxide since upper and lower case letters are not distinguished. The optional entry in the next line shows in parentheses the temperature (K), the pressure (atm.) and the phases of reactants and product in that order. The product temperature "T" is assigned values by the user in the table that is automatically drawn up after the reaction has been entered and the data retrieved. The output is interrupted at the first "?". The user now supplies a numerical value to "T" to compute the extensive state property changes. This is illustrated in the first line for the isothermal reaction at 298 K. In response to the second "T", the user has entered the "DELTA H" value instead of specifying "T". The placement of the asterisk indicates that -200000 J applies to the enthalpy change (heat loss) column. The calculated "T" and associated property changes now refer to a complete (non isothermal) reaction wherein the product carbon dioxide temperature (only) is 1879 K.

The REACTION program permits the entry of up to twelve reactants and twelve products. Provision exists to assign values to pressure, mole numbers and activities in a manner similar to that illustrated here for the temperature. This makes the program suitable for use in many common types of equilibrium calculations. Users may specify before each equation their choice of energy unit (J or cal) and which file(s) are to be used (main file only, their private file only, or the main file superceded/augmented by the contents of their own file). The tasks of examining the data files and entry of user-supplied data are accomplished with other programs, INSPECT and DATAENTRY respectively (3).

EQUILIB

Much of the utility of REACTION depends upon the user's prior knowledge of the products of a reaction. In the case of the example shown in Figure 2, practical heat calculations depend on knowing that carbon dioxide is the main reaction product. The EQUILIB program (4,5) does not require such an assumption. Instead, as shown in Figure 3, the user simply enters the reactants to the equation. Typically, he might wish to alter the C*O to O2 ratio without retyping the reactants each time and so, in this case, the number of moles of O2 has been entered as a variable <A>. The remainder of the user-supplied entry indicated by the line following each "?" shown at the extreme left identifies the reactant temperature (pressure of 1 atm. and most stable phase assumed by default), the value of <A> and the product temperature and pressure. Equilibrium is assumed by the program for the product conditions. All of the possible chemical compounds of C and O found in the main data file and the user's file are listed in the lower half of Figure 3. The user-supplied directives end with a selection of species to be considered in the equilibrium computation. Unless specified to the contrary, the gases are considered in ideal solution and the condensed species as pure phases. Computing time can be saved by not selecting unstable species such as diamond. This selection feature also has the advantage of permitting the study of metastable equilibria in which stable phases are suppressed by omitting them.

Figure 4 shows the computer-generated output. The total number of moles of gas (1.004) is factored from the gas phase so that the mole numbers (fractions) to the left of each species may be immediately interpreted as concentration or partial pressure when, as in the present case, the total pressure is one atmosphere. The zero moles of graphite indicates that this phase was considered in the equilibrium computation but that no graphite was formed. The remainder of the output provides extensive property changes which are similar to that shown in Figure 2. The characters "--- G G -----" immediately above the numbers in the Table at the bottom of Figure 4 indicate that the most stable phases of reactant C*O and O2 are gases.

EQUILIB is very versatile. It can consider the equilibrium reached when up to 20 compounds containing up to 8 different elements react to produce (at equilibrium) up to 99 species in up to 20 potential phases. The Gibbs Phase Rule is observed since the actual number of phases listed with non-zero mole numbers cannot exceed the number of different elements in the system. Types of phases that can be considered include ideal aqueous solutions. The option to consider such solutions is automatically presented when H and O are reactant elements and the temperature and pressure of the products are near the water dew point.

EPH/PREDOM

The previously described F*A*C*T programs, in which users interact with the compound database in a "reaction" mode, are supplemented by programs that provide graphical output in the form of stability diagrams of two standard types... the Pourbaix or redox potential - pH diagram for aqueous systems (EPH) and the isothermal predominance diagram (PREDOM) for non-aqueous systems. Both styles of diagrams have found widespread application in extractive metallurgy, corrosion, geochemistry, and electrochemistry.

Figures 5 and 6 show typical F*A*C*T-generated output. The E-pH diagram in Figure 5 is similar to the diagrams that may be found in many references (6,7). The lines separating the aqueous fields from the oxides and metal field represent the lines shown on a conventional diagram at an activity level of 1E-6 but this level may be changed by the user to suit his application. The valuable features of this program, however, are that users may change the temperature and introduce more than one element beyond the H and O that are intrinsic to the diagram. Coupled with the ability to interface with one's own data file and the provision to introduce metastable phases, the very large number of diagrams that can be produced with the present data base of over 5300 compounds and aqueous ions precludes printing and publication in the usual way.

The PREDOM program is the analog of the Pourbaix diagram. Such a diagram is shown in Figure 6 for the Cu-S-O-Cl system. It shows the conditions of chemical potential associated with the formation of copper-containing compounds. Users are able to select the variables for each axis which may be the logarithms of any combination of partial pressures, activities or ratios thereof that will define the chemical potentials of all elements at any point on the diagram. When, as in the case in Figure 6, more than three elements are involved, it is necessary to fix an additional partial pressure or activity. Failure to correctly specify partial pressures or activity will result in error messages which prevent erroneous output. The diagram in Figure 6 is valuable in indicating stable combinations of phases that contain Cu.

The complications in both the EPH and PREDOM programs resulting from the limited graphics capability of a conventional teleprinter-type terminal are overcome by the option to print the co-ordinates for all triple points and alter the field of view so as to magnify difficult to interpret features (3).

POTCOMP

All of the previously mentioned programs are intended for use with a compounds database in which data on enthalpy of formation, entropy, and heat capacity exist for each stoichiometric phase. The POTCOMP program does not have an interface with this data base. Instead, it is intended for the construction of equilibrium diagrams which usually involve at least some phases of variable composition.

Figure 7 illustrates the relationship between Gibbs energy isotherms and phase boundaries in a simple but practically important type of two phase system. The common-tangent construction in the upper half of the figure is a graphical way of indicating the basic condition for coexisting phases... equal partial molar Gibbs energy for each component in each phase (8). The POTCOMP program (9) allows users to enter constants, a,b,..., in the equation

$$\Delta G = a + bT + cT^2 + dT^3 + eT^4 \qquad (1)$$
$$+ fT \ln T + g/T$$

which define the differences in the Gibbs energy between the different phases of each pure component. Of more interest, POTCOMP also allows for the introduction of excess Gibbs energy of mixing equations for each phase in the form

$$G^E = \sum_{ij} Q_{ij} x_A^i x_B^j \qquad (2)$$

where the coefficients Q have the same form as equation 1. The series may contain up to ten combinations of i and j. These excess functions, when added to the usual expression for ideal mixing, express the degree of droop of each Gibbs energy isotherm for each phase.

The action of the POTCOMP program is to determine the common tangent points (conjugate phase compositions) for a user-specified sequence of temperatures. Output such as that shown in Figure 8 in which the "." fills the vapour field and the "*" fills the liquid field suffice to establish the phase boundaries. Moreover, the option exists to print a table of phase boundary compositions computed to within 0.01 mole percent. An option in the program allows users to construct isothermal pressure-composition diagrams such as that shown in Figure 9. In this case, the relative positions of the Gibbs energy isotherms shift due to the sensitivity of the vapour phase isotherm to pressure.

POTCOMP has the ability to deal with up to seven non-ideal phases. This makes it suitable for calculation of the more complex diagrams involving solid solution phases of interest to materials engineers. Users may permanently store the constants in private data files to speed reconstruction of the diagram at a later date for different conditions of temperature, pressure, and composition range. In addition, all users can share a datafile on several hundred systems that the authors have analyzed. This includes molten chloride systems, alloy systems, ceramic systems and many miscible organic systems such as that shown in Figure 8.

Unlike the other F*A*C*T programs, special versions of POTCOMP are available for those interested in installations on their own mainframe or IBM personal computer.

FITBIN

The program FITBIN is complementary to POTCOMP in that users may enter the co-ordinates of a phase boundary and other thermodynamic information such as enthalpy of mixing and then calculate by regression methods the "best" set of self-consistent constants in equation 2 for a specified phase. These may be automatically stored in the user's private data file to enable him to use POTCOMP subsequently to generate the resulting phase diagram. This FITBIN/POTCOMP cycle was found to be particularly useful in the development of the common POTCOMP datafile to which reference was made in the previous section.

TERNFIG

The program TERNFIG is a three-component version of POTCOMP in which users supply data which define Gibbs energy surfaces for the phases in the system (10). In the absence of data on ternary solutions, an option can be selected to provide a reasonable estimate using one of two proven interpolation techniques (11). This guarantees a feasible first approximation of the phase diagram that can serve as a guide for experimental studies in critical areas. The user may display the liquidus surface projection or an isothermal section. Figure 10 shows a typical liquidus projection diagram.

CONCLUSION

F*A*C*T has been in operation for the past four years and presently serves a user group that includes industrial and governmental labs as well as several universities in both Canada and the United States. Among the features of the system mainly responsible for this growth are the self-consistency of diverse calculations, ease of use by non-computer-trained personnel, low operating cost and the existence of organizational structures which permit continuance of effort in the form of database development and program improvement.

ACKNOWLEDGEMENTS

The authors wish to thank the Natural Sciences and Engineering Research Council of Canada for financial assistance in the form of CO-OP and Strategic grants. They also are grateful to the Computing Centres of McGill University and Ecole Polytechnique for their technical support.

REFERENCES

1. Thompson, W.T., Bale, C.W., and Pelton, A.D., Journal of Metals 32, 18-22 (1980).

2. Thompson, W.T., Pelton, A.D. and Bale, C.W., Engineering Education, 201-205, (1979).

3. Bale, C.W., Pelton, A.D. and Thompson, W.T., "F*A*C*T User's Guide", McGill University/Ecole Polytechnique, Montreal, (1979).

4. Thompson, W.T., Pelton, A.D. and Bale, C.W., CALPHAD 7, 113-123, (1983).

5. Bale, C.W., Pleton, A.D. and Thompson, W.T., "EQUILIB-User's Guide Supplement", McGill University/Ecole Polytechnique, Montreal, (1980).

6. Garrels, R.M. and Christ, C.L., "Minerals, Solutions and Equilibria", Harper and Row, New York, (1965).

7. Pourbaix, M., "Atlas of Electrochemical Equilibria in Aqueous Solutions", Pergamon, London, (1963).

8. Pelton, A.D. and Thompson, W.T., Progress in Solid State Chemistry, 10, 119-155, (1975).

9. Bale, C.W., Pelton, A.D. and Thompson, W.T., "POTCOMP-User's Guide Supplement", McGill University/Ecole Polytechnique, Montreal, (1981).

10. Lin, P.L., Pelton, A.D., Bale, C.W. and Thompson, W.T., CALPHAD, 4, 47-60, (1980).

11. Lin, P.L., Pelton, A.D. and Bale, C.W., J. Amer. Ceram. Soc. 62, 414-424, (1979).

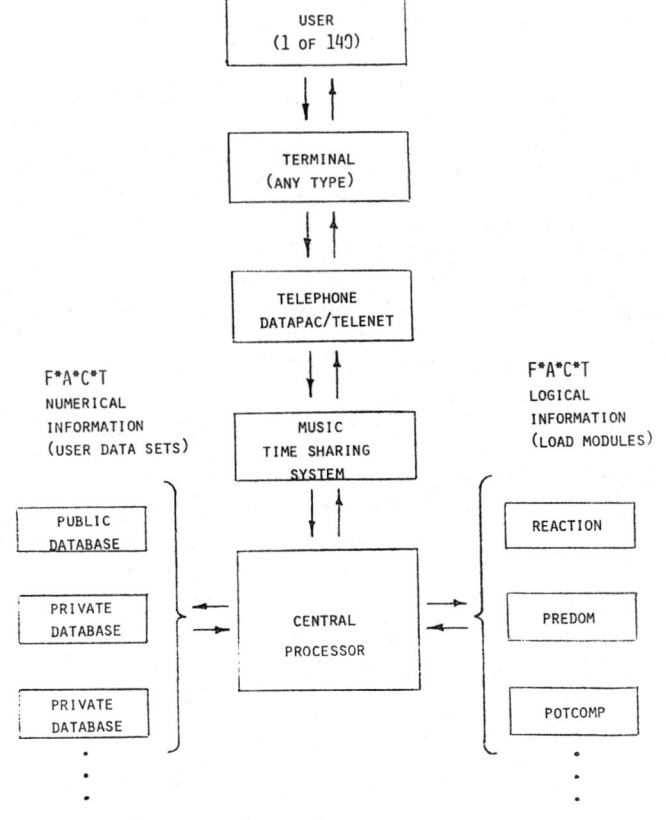

Figure 1. F*A*C*T system architecture.

```
XXXXXXXXXXXXXXXXXXXXXXXXXXXXXXXXXXXXXXXXXXXXXXXXXXXXXXXXXXXXXXXXXXXXXXXXX

     C*O        +       1/2 O2        =        C*O2
   (298,1,GAS)         (298,1,GAS)           (T,1,GAS)

          CALCULATIONS ARE BASED ON THE INDICATED NUMBER OF GRAM MOLES

*************************************************************************
    (T)      DELTA H      DELTA G      DELTA V     DELTA S      DELTA U      DELTA A
    (K)       ( J )        ( J )         (L)       ( J /K)       ( J )        ( J )
*************************************************************************
?
 298
    298.0    -282962.3    -257213.1    -0.122E+02   -86.406     -281723.1    -255974.0
?
  *          -20000
   4370.0     -20000.0   -1522097.0    0.322E+03    64.148      -52620.4    -1554717.0
?
```

Figure 2. Calculation of extensive state function changes for the stoichiometric combustion of carbon monoxide using the REACTION program. The first line in the table gives the property changes for the isothermal reaction. The second line gives the property changes for the case where there is a heat loss of 200000 J/mol carbon dioxide.

```
****** ENTER REACTANTS ******(OR PRESS "RETURN" FOR LAST ENTRY)
?
     C*O     +    <A> O2    =    ?
ENTER NEW SUBSCRIPTS, TYPE "R", ENTER "HELP" OR PRESS "RETURN"
?
     (298)        (298)
SPECIFY A VALUE FOR "A"
?
0.5
SPECIFY THE PRODUCT TEMPERATURE (K)
?
1879
SPECIFY THE PRODUCT PRESSURE (ATM) OR PRESS "RETURN" FOR 1 ATM.
(NEGATIVE NUMBER TREATED AS PRODUCT GAS VOLUME IN LITERS)
?
1
THERE MAY BE A DELAY WHILE THE LOCATION OF
ALL DATA ON COMPOUNDS CONTAINING THE REACTANT ELEMENTS
IS DETERMINED. THESE WILL BE RECORDED FOR SUBSEQUENT USE.
YOU DO(1) OR YOU DO NOT(2) WISH TO SEE THE LIST OF POSSIBLE PRODUCTS
?
1
POSSIBLE PRODUCT COMPOUNDS FOUND ("<---" WILL IDENTIFY YOUR PRIVATE DATA)
    1   C3O2              G1    GAS        298.0 K -  2000.0 K
    2   C*O2              G1    GAS        298.0 K -  2500.0 K
    3   C*O               G1    GAS        298.0 K -  2500.0 K
    4   O3                G1    GAS        298.0 K -  2000.0 K
    5   O2                G1    GAS        298.0 K -  3000.0 K
    6   O                 G1    GAS        298.0 K -  2000.0 K
    7   C5                G1    GAS        298.0 K -  2000.0 K
    8   C4                G1    GAS        298.0 K -  2000.0 K
    9   C3                G1    GAS        298.0 K -  2000.0 K
   10   C2                G1    GAS        298.0 K -  4000.0 K
   11   C                 G1    GAS        298.0 K -  2000.0 K
   12   C                 S1    GRAPHITE  1100.0 K -  4073.0 K
   13   C                 S2    DIAMOND    298.0 K -  1200.0 K
ENTER CODE.NUMBERS OF SPECIES TO BE CONSIDERED OR ENTER "HELP"
?
1-12
PRESS "RETURN" WHEN READY FOR OUTPUT
?
```

Figure 3. Input to the EQUILIB program to determine the products of carbon monoxide oxidation at 1879 K.

```
    C*O      +    <A> O2     =
   (298)           (298)


                         1.004   {     .989         C*O2
                                 +     .743E-02     C*O
                                 +     .370E-02     O2
                                 +     .152E-04     O
                                 +     .688E-11     O3
                                 +     .486E-20     C
                                 +     .285E-22     C3O2
                                 +     .590E-30     C2
                                 +     .337E-37     C3
                                 +     .303E-50     C4
                                 +     .138E-58     C5                        }
                                                  (1879.0,  1.00      ,G  )

                                 +    .0            C
                                                  (1879.0,  1.00      ,S1)


WHERE 'A' ON THE REACTANT SIDE IS          0.5000000E+00

UNITS: JOULES(1) OR CALORIES(2)?
?
1
           CALCULATIONS BASED ON INDICATED NUMBER OF GRAM MOLES

*******************************************************************************
     DELTA H      DELTA G      DELTA V       DELTA S      DELTA U      DELTA A     REACT V
      ( J )        ( J )        (L)          ( J /K)       ( J )        ( J )       (L)
*******************************************************************************
---  G  G  ----------------------------------------------------------- = AS ABOVE ---
   -197952.4    -683002.4    0.118E+03       5.682      -209916.7    -694966.8   0.367E+02

****** ENTER REACTANTS ******(OR PRESS 'RETURN' FOR LAST ENTRY)
?
```

Figure 4. Output associated with Figure 3 showing the equilibrium composition and the related extensive property changes. Compare with Figure 2.

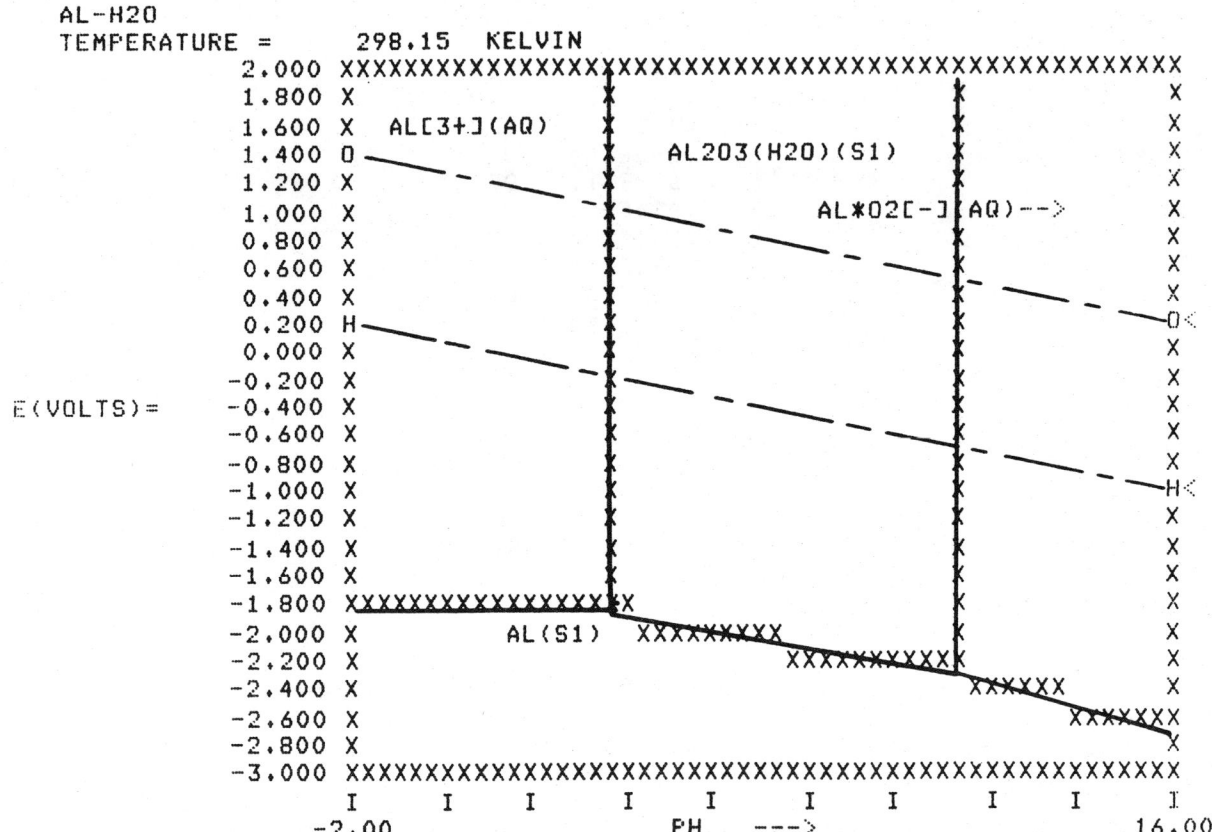

Figure 5. A Pourbaix diagram for aluminum with ionic activities set at 1e-6 and temperature set at 298 K.

```
    CU-S-O-CL
    TEMPERATURE =      1000.00   KELVIN
    CONSTANT LOG10(Z) =    -4.000,   Z = P(S*O*CL2)
    Y = P(S*O3)/P(S*O2)
                 0.0    XXXXXXXXXXXXXXXXXXXXXXXXXXXXXXXXXXXXXXXXXXXX
                -0.500  X                                            X
                -1.000  X                    CU*S*O4(S1)             X
                -1.500  X                                         XXXX
                -2.000  X                                    XXXXX    X
                -2.500  X                                XXXXX        X
                -3.000  X                            XXXXX  CU*CL(L)  X
                -3.500  X                         XXXXX               X
                -4.000  X                      XXXXX                  X
                -4.500  XXXXXXXXXXXXXXXXXXXXXXXX                      X
                -5.000  X           X   CU2S(S3)  XX                  X
                -5.500  X          XXX             XX                 X
LOG10(Y)=       -6.000  X  CU*S(S1) XX              XX                X
                -6.500  X            XXX             XX               X
                -7.000  X              XX             XX              X
                -7.500  X               XX             XX             X
                -8.000  X                XX             XX            X
                -8.500  X                 XX             XX           X
                -9.000  X                  XX             XX          X
                -9.500  X                   XXXX           XX         X
               -10.000  X                      XXX          XX        X
               -10.500  X                                    XX       X
               -11.000  X                                     XX      X
               -11.500  X                                      XX     X
               -12.000  XXXXXXXXXXXXXXXXXXXXXXXXXXXXXXXXXXXXXXXXXXXXXX
                        I   I   I   I   I   I   I   I   I   I   I
                      -10.00              LOG10(X)--->              0.0
                                        X = P(CL2)


YOU (1) WANT OR (2) NOT WANT A LIST OF INVARIANT POINTS
AND THE SPECIES (IF ANY) THAT ARE NOT LABELLED IN THE DIAGRAM
?
1

INVARIANT POINTS (WITHIN THE RANGE OF AXES -200 TO +200):
LOG10(Y) LOG10(X)         PHASE 1              PHASE 2              PHASE 3
  -4.014   -5.780      4 CU*CL(L)           8 CU*S*O4(S1)        11 CU2S(S3)
 -10.522   -2.526      4 CU*CL(L)          11 CU2S(S3)           13 CU*S(S1)
  -4.421   -8.627      8 CU*S*O4(S1)       11 CU2S(S3)           13 CU*S(S1)

****************************************************************************
DO YOU WISH TO   (1) CONTINUE THE COMPUTATION WITH   CU-S-O-CL
(2) CHANGE TO A DIFFERENT SYSTEM, OR (3) TERMINATE THE PROGRAM
?
```

Figure 6. PREDOM diagram fro the Cu-S-O-Cl system at 1000 K illustrating the domains of stability of all copper-containing compounds in gas mixtures of S*O3, S*O2, S*O*CL2 and CL2.

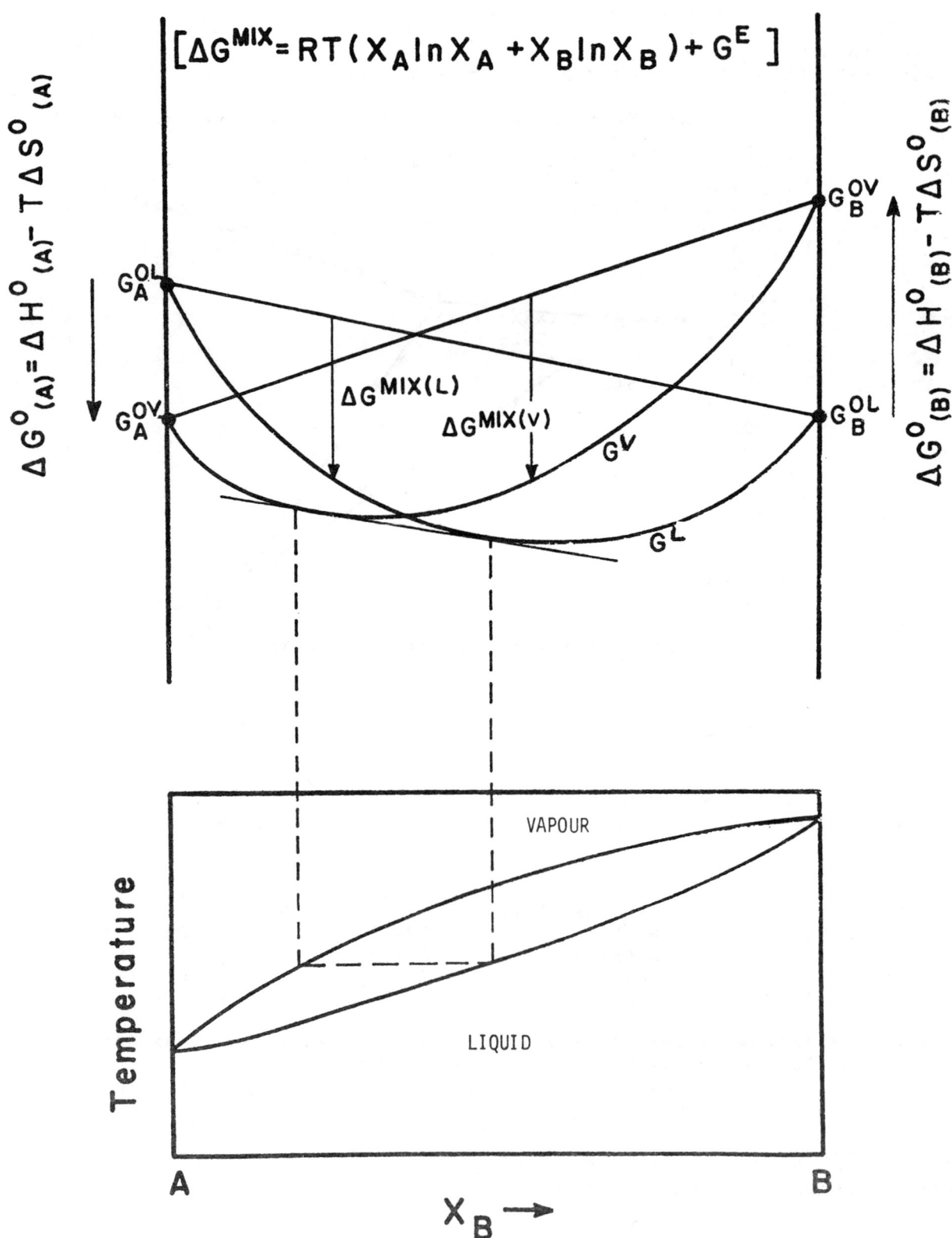

Figure 7. Relationship between Gibbs energy isotherms and phase boundaries in a two-phase system.

```
SYSTEM A-B

A = BENZENE
B = TOLUENE
BENZENE + TOLUENE VLE

PHASES

  *   LIQUID (RECOMMENDED 1 ATM)
  .   GAS

P(TOTAL) = CONSTANT =    1.000E+00

                    BENZENE           MOLE FRACTION -->           TOLUENE

                 0.00      0.20     0.40     0.60     0.80      1.00
         TEMP (C) I    I    I    I    I    I    I    I    I    I    I

         118.000  ................................................
         116.000  ................................................
         114.000  ................................................
         112.000  ................................................
         110.000  ................................................*
         108.000  ..........................................***
         106.000  ............ VAPOUR ..................******
         104.000  ...................................*********
         102.000  .................................***********
         100.000  ...............................*************
          98.000  .............................***************
          96.000  ...........................*****************
          94.000  .........................*******************
          92.000  .......................*********************
          90.000  ....................************  *************
          88.000  .................************  LIQUID **********
          86.000  ..............************        **********
          84.000  ....*******************************************
          82.000  ..***********************************************
          80.000  **************************************************
          78.000  **************************************************
          76.000  **************************************************

                  I    I    I    I    I    I    I    I    I    I    I
```

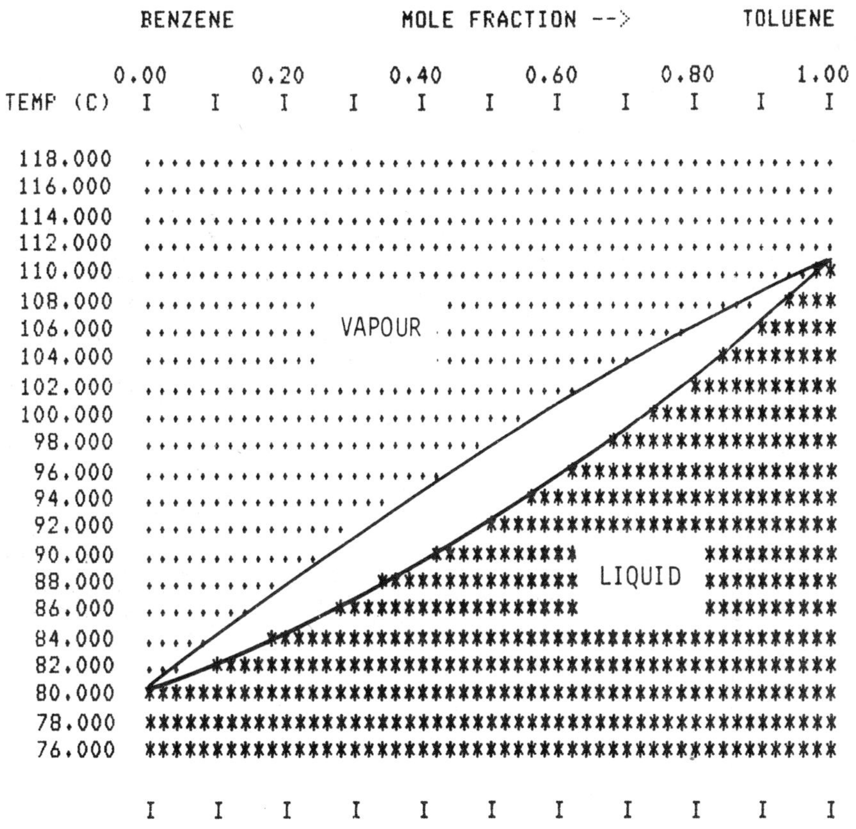

Figure 8. POTCOMP-generated temperature-composition phase diagram at a total pressure of 1 atm. for the benzene-toluene system. Diagram is derived from Gibbs energy equations for each phase.

```
SYSTEM A-B

A = BENZENE
B = TOLUENE
BENZENE + TOLUENE VLE

PHASES

  *    LIQUID (RECOMMENDED 1 ATM)
  .    GAS

TEMP = CONSTANT =    100.00C
```

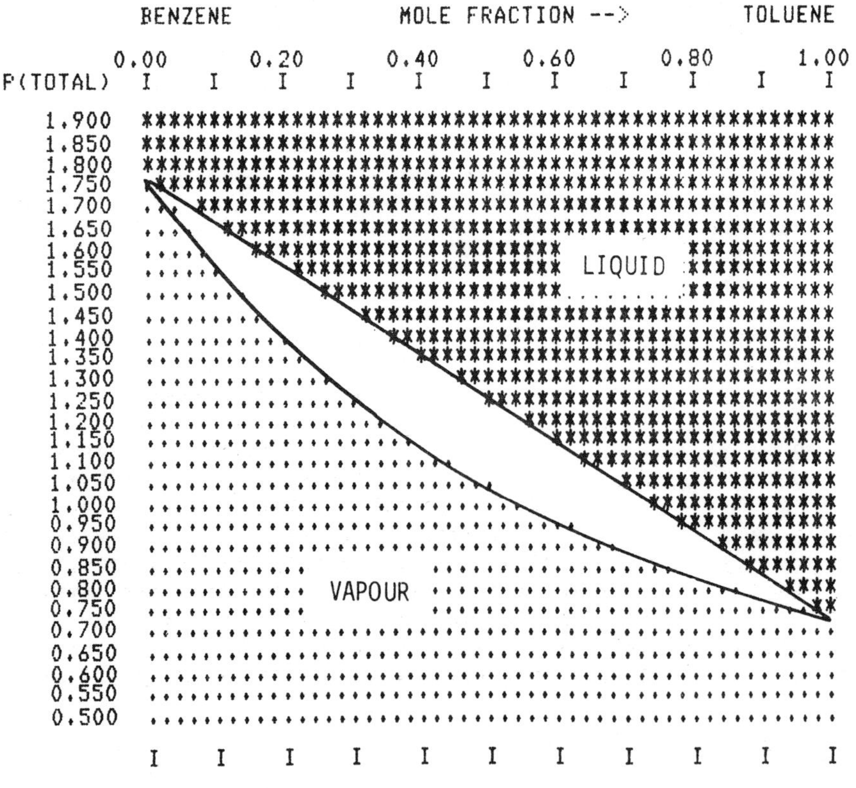

Figure 9. POTCOMP-generated pressure-composition diagram at a fixed temperature of 100 °C. Diagram is calculated from exactly the same data used for Figure 8.

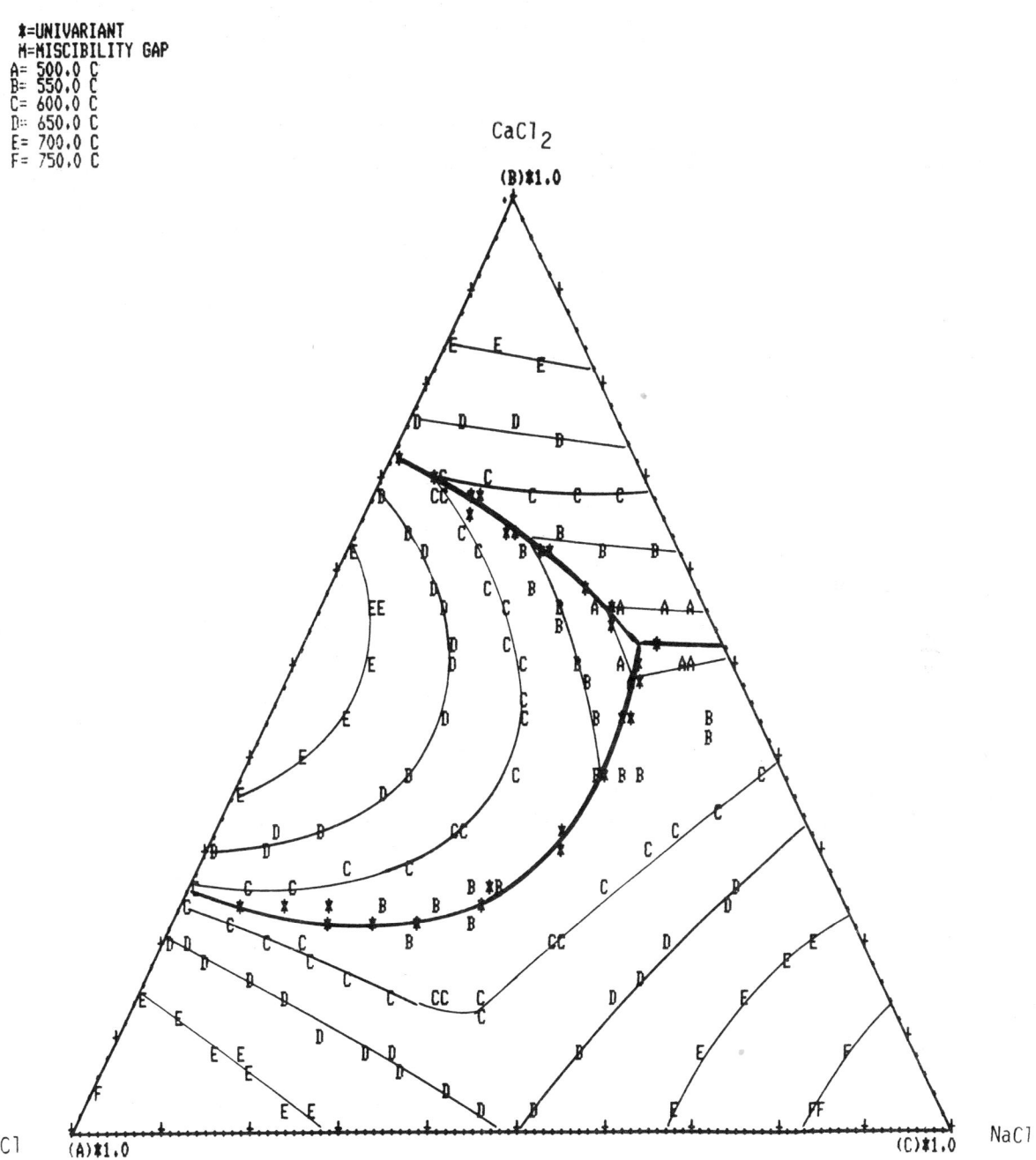

Figure 10. TERNFIG generated liquidus surface projection of the system K*CL-CA*CL2-NA*CL. Diagram was derived from Gibbs energy equations for each phase.

THERMODYNAMIC DATA GENERATION, ESTIMATION, AND COMPILATION FOR MINERAL TECHNOLOGY

N.A. Gokcen ■ Albany Research Center, Bureau of Mines, U.S. Dept. of the Interior, Albany, OR 97321

The Bureau of Mines effort in generating, estimating, and compiling thermodynamic data for minerals and ionic solutions is reviewed. Six apparatuses for calorimetry, one for electrochemistry, and one for ionic activity are briefly described. Recent compilation efforts on thermodynamic properties of the elements, oxides, halides, sulfides, sulfates, carbides, carbonates, nitrides, and hydrides are discussed. Methods for estimation of enthalpies and entropies of formation of alloys are summarized and critically evaluated.

The Bureau of Mines effort in generating, estimating, and compiling thermodynamic data for minerals and ionic solutions is reviewed. Six apparatuses for calorimetry, one for electrochemistry, and one for ionic activity are briefly described. Recent compilation efforts on thermodynamic properties of the elements, oxides, halides, sulfides, sulfates, carbides, carbonates, nitrides, and hydrides are discussed. Methods for estimation of enthalpies and entropies of formation of alloys are summarized and critically evaluated.

The Bureau of Mines has maintained a laboratory for thermodynamic research on minerals and inorganic compounds since 1916. The main objectives of this research effort are to generate and disseminate new thermodynamic data; to evaluate, select, estimate, and compile thermodynamic data that can be used as guidelines for devising efficient processes in mineral technology; and to maintain an up-to-date thermodynamic data bank for providing assistance to technical and research centers. A limited effort is also in progress to generate and to compile ionic activity data useful for leaching, pickling, corrosion prevention, and pollution abatement.

The laboratory for thermodynamic research is integrated to generate accurate data for heat capacities, relative enthalpies, entropies, and enthalpies of formation of a mineral or a synthetic inorganic compound. In addition, a new apparatus has been designed and operated to obtain data on ionic activities of metal salts in water. All the apparatuses used for these investigations are described elsewhere in detail (1-4); hence, only a brief review of the capabilities of each apparatus is presented in the succeeding sections. Further, a bibliography of publications resulting from this effort during the past 5 years is appended for convenience to the reader interested in extensive details of current research efforts.

EXPERIMENTAL INVESTIGATIONS

Low-temperature Calorimeter

The low-temperature calorimeter is a fully automated adiabatic type apparatus that operates over a temperature range of 5 to 300 K. This apparatus directly measures the heat capacity of condensed materials. From the heat capacity measurements, the enthalpy relative to 0.0 K, $H°_T - H°_0$, and entropy $S°_T$ at all temperatures up to 300 K are calculated (1, 5). The calorimeter is operated by first loading the sample into the cryostat, cooling it down by diathermic contact with a liquid helium tank, then lowering the sample in the adiabatic zone, and heating it to room temperature in a stepwise manner under computer control while measuring the temperature and the energy required for heating. The heat capacity is then calculated as the ratio of a small measured energy input to the small temperature rise. A typical set of data is shown in Figure 1.

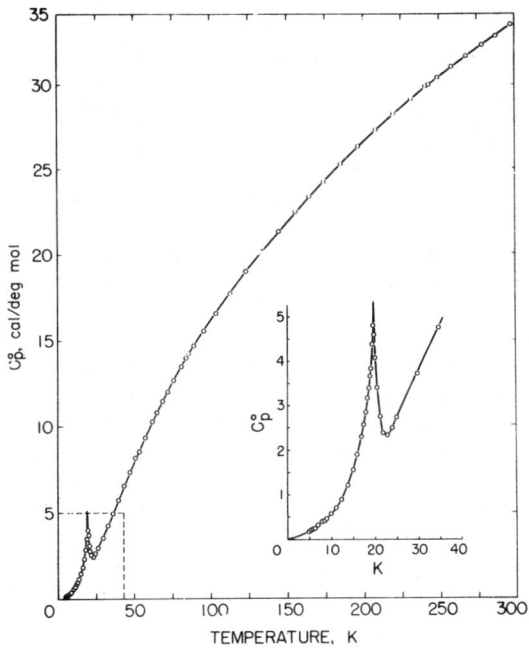

Figure 1. Heat capacity of $CuO \cdot CuSO_4$ at low-temperatures

Compounds for which the heat capacities were measured during the last 5 years have included titanium oxides and sulfides, transition metal oxysulfates, manganese silicates and transition metal halides.

High-temperature Drop Calorimeter

The high-temperature drop calorimeter consists of a copper block within an isothermal jacket and two furnaces. One furnace is for heating the samples up to 1200 K, and the other, for 1200 to 1800 K. A well-characterized sample of known weight is placed in an inert sealed container, usually made of silica or platinum, heated to a desired temperature in one of the furnaces, and then dropped into the copper block of known heat capacity. The rise in the temperature of copper block determines the relative enthalpy of the sample, $H_T^\circ - H_{298}^\circ$. This apparatus was originally described by Douglas and King (6), and subsequently modified and substantially improved as described elsewhere in detail (7). A typical set of data is shown in Figure 2 for Cu_2S and CuS wherein $(H_T^\circ - H_{298}^\circ)/$ (T-298.15) is plotted versus T. The measured relative enthalpies are computer fitted with polynomials in terms of T and the results are used to calculate the heat capacities, entropy increments, and relative enthalpies.

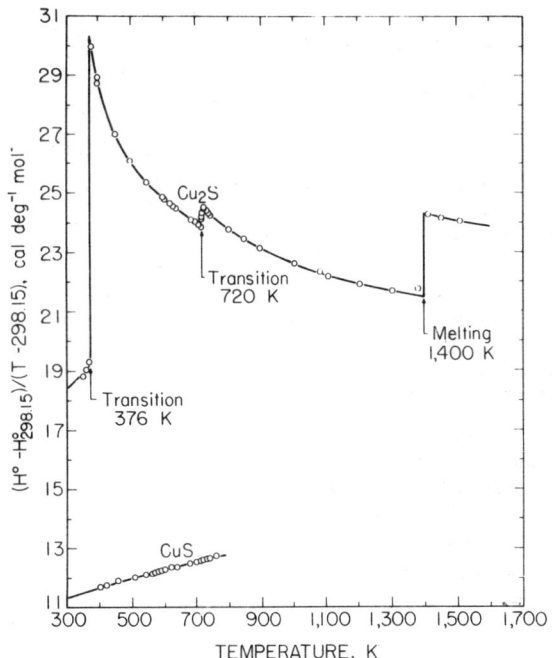

Figure 2. High temperature standard relative enthalpies of CuS and Cu_2S

Differential Scanning Calorimeter

A differential scanning calorimeter (DSC) of the type designated as Mettler TA 2000C* is used for obtaining preliminary heat capacity data for inorganic compounds and minerals with an accuracy of ±2 pct, and ascertaining the first and second order transition enthalpies and temperatures. The same apparatus is also capable of yielding simultaneous thermogravimetric analysis (TGA) with a sensitivity of ±10 micrograms. The range of temperature extends from 25 to 1200° C, and the sample weight ranges from 10 to 6000 mg contained in alumina, platinum or other suitable crucibles. The apparatus operates under vacuum or controlled atmospheres. The rates of heating or cooling are microprocessor controlled with scanning rates of 0.1 to 30° C/min in steps of 0.1° C/min (8). The appar-

---

*Reference in the text to specific trade names does not imply endorsement by the Bureau of Mines.

atus is interfaced to a dedicated microcomputer for instrument control and electronic data processing. The data for transition enthalpies from this apparatus are generally accurate to ±2 pct, and in some cases, more accurate than the data from the high-temperature drop calorimeter. Further, when the DSC shows that there are no solid-solid transitions, fewer measurements are sufficient with the high-temperature drop calorimeter.

This apparatus is also used in characterization of samples synthesized for thermodynamic measurements, and in determinations of approximate vaporization pressures. With controlled atmospheres such as $CO + CO_2$, TGA can be used in chemical equilibria such as $Cu_2O + CO = 2Cu + CO_2$. The apparatus is also useful for phase equilibrium determinations in binary and multicomponent alloys (8).

Solution Calorimetry

Solution calorimetry is a well known technique for determination of the enthalpies of formation for a wide range of minerals and synthetic inorganic compounds (1). Presently the Albany Research Center of the Bureau of Mines has three types of solution calorimeters. The first of these is a constant temperature jacket type that uses a 2-liter glass Dewar reaction vessel. The instrumentation for measurement of temperatures and energy calibrations of an earlier design by Southard (9) was modernized, using recent technical innovations. The solvent used in this apparatus is 4.360 molal HCl, which is sometimes amended with an oxidizer such as ceric sulfate or potassium dichromate to facilitate conversion of multivalent solutes to stable end products. Dissolution measurements are generally conducted at 298.15 K with the sample sealed in a fragile glass ampoule submerged in the solvent prior to initiation of the reaction. The high resolution measurement system uses a slightly off balance resistance bridge as a thermometer in conjunction with a sub-microvolt potentiometric detector. The overall temperature resolution of this arrangment is about ±25 microkelvins in reaction temperature increments of several tenths of degrees. Energy calibration of the calorimetric system is accomplished after the dissolution period by electrical means using a calibrated resistor built into the calorimeter.

Derivation of the enthalpy of formation generally requires a sequence of solution reactions to obtain the necessary data as in the following reaction and equation:

$$AB+CD = AD+CB \quad (1)$$

$$\Delta H_1 = \Delta Hf°(AD)+\Delta Hf°(CB)-\Delta Hf°(AB)-\Delta Hf°(CD) \quad (2)$$

If ΔH_1, the enthalpy of reaction (1) is determined by solution calorimetry, and the corresponding enthalpies of formation of three other components are either known or also determined, then the $\Delta Hf°$ (AB) for the formation of AB can be obtained. Often several dissolution processes are required, leading to a complex reaction scheme so that $\Delta Hf°$ (AB) can be calculated.

The second reaction calorimeter is an HF solution calorimeter. It is primarily used to obtain enthalpies of formation of silicate minerals, and occasionally, refractory materials insoluble in other inorganic acids. This apparatus has also undergone considerable improvement since its initial construction in 1948. The calorimeter reaction vessel is made of platinum -10% rhodium alloy and contains about 1 liter of 20 wt % HF solution held at 73° C for dissolving the silicates. This calorimeter also uses an isothermal jacket. The primary use of this calorimeter is the determination of formation enthalpies of natural and synthetic silicate minerals. The complexity of these materials present formidable analytical and experimental problems. Further, several dissolution processes leading to a complex reaction scheme are required to obtain $\Delta H_f°$ for a selected silicate mineral.

The third reaction calorimeter is the acid bromine calorimeter, recently developed to obtain the enthalpy of formation of sulfide minerals not amenable to conventional acidic solvents. This calorimeter uses a nearly saturated bromine solution in 2.5 molal HBr and is capable of dissolving a wide range of transition metal sulfides and intermetallic compounds. The reaction vessel is fabricated from a tantalum alloy and holds about 700 ml of solvent. It is an adiabatic calorimeter designed to track accurately long reaction periods of 2 to 5 hours required to dissolve a typical refractory sulfide. The reaction vessel is hermetically sealed and operates up to 50° C, about 8° below the normal boiling point of bromine. Data acquisition is fully automated by using a small desktop microcomputer and high-resolution digital instrumentation.

The reaction scheme for a sulfide used in this calorimeter is much simpler than that for the other calorimeters, e.g.,

$$S(\text{rhombic}) = S(\text{in soln}); \quad \Delta H_1 \quad (3)$$
$$M(\text{metal}) = M(\text{in soln}); \quad \Delta H_2 \quad (4)$$
$$MS(\text{sulf}) = M(\text{in soln}) + S(\text{in soln}); \quad \Delta H_3 \quad (5)$$
$$M + S = MS(\text{sulf}); \quad \Delta H_f^\circ = \Delta H_1 + \Delta H_2 - \Delta H_3 \quad (6)$$

Thus the standard enthalpy of formation of a sulfide, MS, is equal to the algebraic sum of ΔH_i, (i=1,2,3) for three simple reactions, provided that the concentrations of M and S in solution are the same in all three reactions in the same type of solutions (1).

Galvanic Cells

The standard Gibbs energies of formation of various compounds are being determined by measuring the electromotive force (emf) of a suitable cell using an appropriate solid electrolyte. Calcium fluoride, doped zirconia, and other similar solids provide convenient media for anionic diffusion and thus generate emf in properly designed galvanic cells (10-12). For example, in the following cell:

$$\text{Pt} | (\text{Fe}+\text{FeF}_2) | \text{CaF}_2(\text{solid electrolyte}) | (\text{Ni} + \text{NiF}_2) | \text{Pt} \quad (7)$$

the measured emf between two Pt terminals is generated by the F^- anion potential difference across the CaF_2 solid electrolyte that permits only the migration of F^- anion. The overall cell reaction and the corresponding $\Delta G°$ are

$$\text{Fe} + \text{NiF}_2 = \text{Ni} + \text{FeF}_2, \quad \Delta G° = -2FE° \quad (8)$$

where F is the Faraday constant and $E°$ is the measured emf. Combination of $\Delta G°$ for this reaction with the known value of the standard Gibbs energy of formation, ΔG_f°, for NiF_2 yields ΔG_f° for FeF_2. (1, 10-12).

Ionic Activities of Metal Salts

The ionic activities of metal salts in water provide scientific and technical bases for leaching, pickling, evaporation, corrosion prevention, and pollution control. The available data on various solutions at various temperatures and pressures are inadequate. The existing data for 25° C have been compiled by Robinson and Stokes (14), Harned and Owen (15); see also Staples (16) and Goldberg (17), as well as the references cited in these publications by these authors. The data for selected electrolytes at various temperatures and molalities are currently being compiled at the Bureau of Mines.

The ionic activities of metal salts in water are being measured with an apparatus patented by the Bureau of Mines (13). This apparatus measures the difference in vapor pressures of pure water and a known solution by using a high-precision differential pressure transducer. The entire assembly is designed to make measurements with five solutions of appropriate concentrations in a single day, each time using one solution against pure water. The method is rapid and highly accurate. For example, if a pressure difference of $\Delta P = 2$ Torr over a solution at 25° C is measured with an accuracy of 0.1%, the activity of water in that solution is determined with an accuracy of 0.009%; the activity of water, a_w, being defined by $a_w = (P° - \Delta P)/P°$ where $P°$ is the vapor pressure of pure water. For large values of ΔP, small fugacity coefficient corrections must be applied to the pressure ratio defining a_w (2, 18). The experimental values of a_w are then fitted to the following polynomial function of molality m:

$$55.508 \ln a_w = -\nu m + Bm^{1.5} + Cm^2 + Dm^{2.5} + \ldots (9)$$

where ν is the total number of ions formed by complete ionization of an electrolyte, B is directly related to the coefficient in the Debye-Hückel limiting equation, and C, D, ... are empirical coefficients. The same coefficients appear in the equation for the activity coefficient γ of an electrolyte as required by the Gibbs-Duhem relation (2, 18):

$$\ln \gamma = -\frac{1.5}{0.5}Bm^{0.5} - \frac{2}{1}Cm - \frac{2.5}{1.5}Dm^{1.5} + \ldots (10)$$

Measurements of this type at various temperatures provide the temperature dependent equations for a_w and γ.

The values of a_w and γ for various solutions are estimated by using the Kusik-Meissner method (2, 19) when the data are not available, particularly for multicomponent solutions. Confidence is gained by also using the methods proposed by Bromley (20) and by Pitzer (21-22), whenever possible. It is the author's experience that the Kusik-Meissner method is the most convenient and comprehensive method (for a summary see ref. 2).

CRITICAL EVALUATION AND COMPILATION OF DATA

The Bureau of Mines critical evaluation and compilation began in Berkeley, California, in 1932, when K. K. Kelley published the first volume of a long series under the title "Contributions to the Data on Theoretical Metallurgy." His first work, subtitled, "The Entropies of Inorganic Substances," (23) was followed the next year by, "High-Temperature Specific Heat Equations for Inorganic Substances," (24). Three more additions to the theoretical metallurgy series were published in the following 3 years: "The Free Energies of Vaporization and Vapor Pressures of Inorganic Substances," (25); "Metal Carbonates—Correlations and Application of Thermodynamic Properties," (26); and "Heats of Fusion of Inorganic Substances," (27).

By 1936, sufficient data had accumulated in the literature for Kelley to prepare a second edition of the first volume of "Contributions to the Data on Theoretical Metallurgy," subtitled, "A Revision of the Entropies of Inorganic Substances," (28), and two more additions to the series appeared in 1937, "The Thermodynamic Properties of Sulfur and Its Inorganic Compounds," (29) and "The Thermodynamic Properties of Metal Carbides and Nitrides," (30).

In the following years, Kelley's compilation work was interrupted, first by experimental low-temperature heat-capacity investigations and then by supervisory duties as chief of the Bureau of Mines thermodynamics group in Berkeley. In spite of added responsibilities, he continued his compilation interests, and in 1941 published a third edition of the popular entropy bulletin, "The Entropies of Inorganic Substances. Revision (1940) of Data and Methods of Calculation," (31).

From 1949 to 1950, after several shorter compilations, Kelley published two more volumes of the theoretical metallurgy series. "High-Temperature Heat-Content, Heat-Capacity and Entropy Data for Inorganic Compounds," (32), was the first revision of the 1933 high-temperature bulletin, and "Entropies of Inorganic Substances. Revision (1948) of Data and Methods of Calculations," (33) was the fourth edition of the widely circulated entropy bulletin.

As the volume of published data expanded at an increasing rate, Kelley recognized the need for skilled evaluators and trained J. P. Coughlin, A. D. Mah, and E. G. King in the critical evaluation and compilation of thermodynamic properties. In 1954, J. P. Coughlin published the 12th volume of the "Contributions to the Data on Theoretical Metallurgy," series, "Heats and Free Energies of Formation of Inorganic Oxides," (34).

In 1955, Kelley wrote a chapter titled, "Thermodynamic Properties of Zirconium Compounds," for the book, "Metallurgy of Zirconium," (35). In 1959, Kelley published with Mah, "Metallurgical Thermochemistry of Titanium," (36).

Three compilations came out in 1960. The foremost was Kelley's third and final revision of the high-temperature work, Bureau of Mines Bulletin 584, "Contributions to the Data on Theoretical Metallurgy. XIII. High-Temperature Heat-Content, Heat-Capacity, and Entropy Data for the Elements and Inorganic Compounds," (37). This valuable work is still in use. Following the high-temperature bulletins came a chapter, "Thermodynamics of Hafnium and Its Compounds," written by Kelley and King for the Naval Reactor Handbook, "The Metallurgy of Hafnium," (38). This was followed by "Thermodynamic Properties of Manganese and Its Compounds," by Mah (39).

In 1961, Kelley and King published the fifth and final revision of the entropy compilation, Bureau of Mines Bulletin 592, "Contributions to the Data on Theoretical Metallurgy. XIV. Entropies of the Elements and Inorganic Compounds," (40). Like Bulletin 584, this work is still in use. In 1962, in response to demand, four of Kelley's older bulletins were republished as Bureau of Mines Bulletin 601, "Contributions to the Data on Theoretical Metallurgy. XV. A Reprint of Bulletins 383, 384, 394, and 404" (41).

After Kelley's retirement, Mah continued the critical compilation work and in 1966 published "Thermodynamic Properties of Vanadium and Its Compounds," (42). After the Berkeley Thermodynamics Laboratory moved to its present location in Albany, Oregon, King's computerization of compilation methodology provided techniques that have been useful up to the present time. Further modernization of the compilation effort has been accomplished after 1977, using up-to-date data processing methods. An in-house minicomputer system designed by J. M. Stuve for this purpose was successfully implemented, with adaptation of software by R. P. Beyer.

In 1973, by cooperative agreement with the International Copper Research Association, King, Mah, and L. B. Pankratz published a monograph, "Thermodynamic Properties of Copper and Its Inorganic Compounds," (43), and in 1976 Mah and Pankratz published, "Contributions to Data on Theoretical Metallurgy. XVI: Thermodynamic Properties of Nickel and Its Inorganic Compounds," (44).

In 1982, Pankratz published Bulletin 672, "Thermodynamic Properties of Elements and Oxides," (45); C. W. DeKock, IC 8910, "Thermodynamic Properties of Selected Transition Metal Sulfates and Their Hydrates," (46); and Mah, RI 8671, "Thermodynamic Data for Arsenic Sulfide Reactions," (47). These last five compilations differ from their predecessors in that they include heat capacities, standard entropies, and enthalpies and Gibbs energies of formation. Current compilation efforts include thermodynamic properties of halides, hydrides, carbides, nitrides, and alkali and alkaline earth sulfates.

Procedure For Data Evaluation

The Bureau of Mines compilation effort has been facilitated by the continuing effort to maintain an up-to-date file of thermodynamic data from the literature. Related data such as vapor pressures, temperatures and enthalpies of transformation, decomposition studies, phase studies, and pertinent gas data are also collected. The major source of references is from a search of the Chemical Abstracts. However, references in collected articles and bibliographies of compilations and review articles are also examined.

After collection, the first step of evaluation is to assess the method used to generate the data. Though the following discussion centers on the copper-block drop calorimeter, it should illustrate the processes used. Particular points of the system examined are the temperature measurement of the sample, calibration techniques, the calorimeter itself, and the actual measurements.

The sample temperature-sensing device, usually a thermocouple, should be referred to some well-defined temperature scale by appropriate calibration. Also, the method of calibration must show any changes after use if the thermocouple was exposed to high temperatures; for example, after use above 1,500 K, platinum versus Pt-Rh couples must be recalibrated. Needless to say, the equipment associated with the thermocouple should not degrade its accuracy. Regardless of the care taken to insure accuracy of measurement, the resulting data would be useless if the sensing device had not indicated the true temperature of the sample through poor positioning, or if part of the sample was hotter or colder than the indicated temperature because of an inadequate isothermal zone in the furnace. Along with the calibration of the temperature-measuring device, an examination is made of the techniques used for other equipment, such as standard cells, standard resistors, timers, and potentiometer.

The calorimeter and the actual measurements are studied from several points. First, is the method appropriate? For instance, drop calorimeters are not very precise below 400 K, nor are they suitable for cases where the sample does not return to the same well defined state upon cooling. Second, are the design and operation such that the heat exchange between the calorimeter and its surroundings is well defined and accounted for? Obviously, any unknown or unaccounted heat leak will affect the accuracy of the measurements. Third, how was the calorimeter calibrated? This must be done by direct electrical measurement and, if done often, will reflect any small, long-term changes. Comparison of measurements of standard substances, such as alpha alumina, with those of other workers can indicate the overall accuracy and precision of the system but should not be used as a calibration.

The second step of evaluation is to assess the sample and its treatment. Two particular areas are examined, characterization of the sample and possible reactions or changes occurring during measurements. Certainly, a careful analysis of the sample composition and stoichiometry should be made. Other data needed for a more complete characterization of the sample would be its crystal structure, the presence or absence of other phases, and the method of preparation. During measurements, some check should be made to determine if the sample is returning to the same state on cooling in the calorimeter. Also, during measurements, the possibilities of decomposition vaporization and reaction with the container or with impurities should be examined. Sometimes a complete reanalysis of the sample is necessary after completion of the measurements. Some final points in assessing the data include corrections for the sample container, for impurities, and for computational procedures.

Except for heat capacities, which are treated separately, all evaluated data are converted to enthalpies relative to 298.15 K in calories per mol and compared with one another. These data are plotted as the function $(H_T^\circ - H_{298}^\circ)/(T-298.15)$ versus T in K. (The subscript 298 is abbreviation of 298.15 K.) This function is used because it emphasizes enthalpy differences; also, it is equal to the heat capacity at 298.15 K. All data for a single substance are plotted on the same graph with the enthalpy data derived from low-temperature data. Only a short range of low-temperature data, usually in the range of 260 to 310 K, is used, since the primary purpose is to achieve a reasonable merging of high and low-temperature data.

The evaluation of the calorimetric systems and samples, the $(H_T^\circ - H_{298}^\circ)/(T-298.15)$ versus T plot, and any other pertinent data, such as phase studies, transformation temperatures and enthalpies, and decomposition studies, are then used to select "the best" values. These selected experimental values are fit with a polynomial in temperature, using a computer program. Several calculations with different parameters are usually needed before an acceptable fit is obtained. The program also calculates and tabulates, at even temperatures and at transition points, heat capacities, entropies, Gibbs energy functions, and relative enthalpies.

Two real problems may remain at this point. The first is when two apparently equally good studies disagree well outside their experimental errors. The resolution of this disagreement requires the complete reassessment of the evaluation and the examination of any other work not previously studied. This examination includes results for different substances by the authors under consideration, as well as those by other investigators. The second problem is the estimation of unavailable data. Though many methods of estimation have been published, none have seemed adequate in all cases. As a consequence, estimations are rarely made, and then only with some supporting evidence and consultation with others.

The collection and evaluation of data for the elements and oxides are published as Bureau of Mines Bulletin 672 in 1982 as mentioned earlier. The tabular data include heat capacities, standard entropies, Gibbs energy functions, and enthalpies relative to 298.15 K. Tables for the oxides also include the enthalpies and Gibbs energies of formation. Standard-form equations have been fit to the tabular relative enthalpies. These equations, along with those derived for heat capacities, heats of formation, and Gibbs energies, are also given.

THERMODYNAMIC PROPERTIES OF ALLOYS

The available data on thermodynamic properties of binary alloys prior to 1974 were evaluated, compiled and published by Hultgren et al (48), with the participation of Kelley. Since that time no comparable effort has been undertaken. Measurements of enthalpies and entropies of formation of alloys are often difficult and much of the available data are inadequate. For unknown data, it is necessary to rely on the best methods of estimation based on the properties of the pure component elements. An empirical method developed by Miedema et al (49-51), is adopted in the author's work on the enthalpy of formation of binary and multicomponent solid and liquid alloys. The method is not too complicated for the alloys of transition elements, but for the non-transition elements, it is fairly complex. The estimation procedure has been summarized and illustrated with numerical examples by the author (52).

The entropies of solution of binary alloys are even more difficult to estimate than the enthalpies of solution. A satisfactory procedure has not yet been developed. As a preliminary method for rough estimations, Kubaschewski (53) suggested that for a dilute solution of a metal, A, in a solvent metal, B, the excess partial molar enthalpy of solution of A, $\bar{H}_A^e$, is related to the excess partial molar entropy of solution of A, $\bar{S}_A^e$, by $\bar{H}_A^e = 3400\, \bar{S}_A^e$, where 3400 is an empirical constant whose dimension is temperature in K. Since $\bar{H}_A^e$ is either known or can be estimated by the Miedema method, it is possible to estimate $\bar{S}_A^e$ when data are not available.

The foregoing methods can be refined as more experimental results become available. The acid bromine calorimeter is useful for determination of ΔH of formation for alloy phases, and the DSC can be adapted for ΔH of formation of liquid alloys by suitable design of crucibles.

BIBLIOGRAPHY

The bibliography of the publications resulting from the Bureau research effort in thermodynamics since 1978 is cited at the end of this paper for the convenience of the reader interested in pursuing the details of experimental work and using the results.

LITERATURE CITED

1. N. A. Gokcen, R. V. Mrazek, and L. B. Pankratz. Proceedings of the Workshop on Techniques for Measurement of Thermodynamic Properties, Albany, Oreg., August 21-23, 1979, BuMines IC 8853 (1981).

2. N. A. Gokcen, Determination and Estimation of Ionic Activities of Metal Salts in Water, BuMines RI 8372 (1979).

3. N. A. Gokcen, "Hydrometallurgy-Research, Development and Plant Practice," edited by K. Osseo-Asare and J. D. Miller, Conference Proc., TMS-AIME (1982).

4. J. P. McCullough and D. W. Scott editors, "Experimental Thermodynamics, Vol. I," Plenum Press New York (1968); [Vol. II, edited by B. Vodar and B. LeNeidre, Pergamon Press, New York (1974)].

5. R. P. Beyer, M. J. Ferrante, and R. V. Mrazek, "An Automated Calorimeter for Heat Capacity Measurements From 5 to 300 K. The Heat Capacity of Cadmium Sulfide From 5.37 to 301.8 K and the Relative Enthalpy to 1103.4 K," accepted for publication in J. Chem. Thermo. (1983).

6. T. B. Douglas and E. G. King, "High-Temperature Drop Calorimetry," Vol. I, Experimental Thermodynamics, J. P. McCullough and D. W. Scott editors, Plenum Press, New York, p. 193 (1968).

7. M. J. Ferrante, "High-Temperature Enthalpy Measurements With A Copper-Block Calorimeter," p. 134 in Proceedings of the Workshop on Techniques for Measurement of Thermodynamic Properties, Albany, Oreg., Aug 21-23, 1979, BuMines IC 8853 (1981).

8. Mettler Instrument Corp, "TA 2000C Thermo-analyzer for Simultaneous TG-DSC," Mettler 1.7300.73A, Box 71, Hightstown, NJ.

9. J. C. Southard, "Heat of Hydration of Calcium Sulfates," Ind. Eng. Chem. $\underline{32}$, 442 (1940).

10. S. C. Schaefer, The Free Energy of Formation of Indium Oxide by EMF Measurements. BuMines RI 7549, (1971).

11. S. C. Schaefer, Free Energies of Formation of Ferrous and Ferric Fluoride by Electromotive Force Measurements. BuMines RI 8096, (1975).

12. S. C. Schaefer, Free Energies of Formation of Chromous, Chromic, and Chromium (I, II) Fluorides by Electromotive Force Measurements. BuMines RI 8172, (1976).

13. N. A. Gokcen, "Method for Measuring the Ionic Activities in Water with Differential Pressure Transducers," U.S. Patent 4,301,676 (1981).

14. R. A. Robinson, and R. H. Stokes, Electrolyte Solutions, Butterworths Scientific Publishers, London, 2d rev. ed., 5th impression (1970).

15. H. S. Harned, and B. B. Owen, The Physical Chemistry of Electrolytic Solutions, Reinhold Publishing Corp., New York, 3d ed. (1958).

16. B. R. Staples, Evaluation of Activity And Osmotic Coefficients For Electrolyte Solutions: Basic Methodology, p. 286 in Proceedings of the Workshop on Techniques for Measurement of Thermodynamic Properties, Albany, Oreg., Aug 21-23, 1979, BuMines IC 8853, compiled by N. A. Gokcen, R. V. Mrazek, and L. B. Pankratz, (1981).

17. R. N. Goldberg, Evaluation Of Activity And Osmotic Coefficients For Electrolyte Solutions: Applications to Real Systems, p. 293 in Proceedings of the Workshop on Techniques for Measurement of Thermodynamic Properties, Albany, Oreg., Aug 21-23, 1979, BuMines IC 8853, compiled by N. A. Gokcen, R. V. Mrazek, and L. B. Pankratz, (1981).

18. N. A. Gokcen, Thermodynamics, Chapters XII and XIV, Techscience, Inc., Hawthorne, Calif. (1975).

19. C. L. Kusik, and H. P. Meissner, "Activity Coefficients in Hydrometallurgy," Presented at 106th AIME Ann. Meeting, Atlanta, Ga., Mar. 8 (1977). Available from C. L. Kusik at Arthur D. Little, Inc., Cambridge, Mass.; "Calculating Activity Coefficients in Hydrometallurgy--A Review," Internat. J. of Miner. Proc., $\underline{2}$ (1975), 105; "Vapor Pressures of Water Over Aqueous Solutions of Strong Electrolytes," I & EC Process Des. and Develop., $\underline{12}$ (1) 112 (1973).

20. L. A. Bromley, "Thermodynamic Properties of Strong Electrolytes in Aqueous Solutions," AIChE J., $\underline{19}$ (2) 313 (1973).

21. K. S. Pitzer, "Thermodynamics of Electrolytes. I. Theoretical Basis and General Equations," J. Phys. Chem., $\underline{77}$ (1973) 268; K. S. Pitzer and G. Mayorga, "Thermodynamics of Electrolytes." II.,

J. Phys. Chem., 77 2300 (1973).

22. K. S. Pitzer, *Activity Coefficients in Electrolyte Solutions*, 1, R. M. Pytkowicz, editor, CRC Press, Boca Raton, Fla. 157 (1979).

23. Kelley, K. K., Contributions to the Data on Theoretical Metallurgy. I. The Entropies of Inorganic Substances, BuMines B 350 (1932).

24. Kelley, K. K., Contributions to the Data on Theoretical Metallurgy. II. High-Temperature Specific Heat Equations for Inorganic Substances. BuMines B 371, (1933).

25. Kelley, K. K., Contributions to the Data on Theoretical Metallurgy. III. The Free Energies of Vaporization and Vapor Pressures of Inorganic Substances. BuMines B 383, (1935).

26. Kelley, K. K., and C. T. Anderson, Contributions to the Data on Theoretical Metallurgy. IV. Metal Carbonates--Correlations and Application of Thermodynamic Properties. BuMines B 384, (1935).

27. Kelley, K. K., Contributions to the Data on Theoretical Metallurgy. V. Heats of Fusion of Inorganic Substances. BuMines B 393, (1936).

28. Kelley, K. K., Contributions to the Data on Theoretical Metallurgy. VI. A Revision of the Entropies of Inorganic Substances. BuMines B 394 (1936).

29. Kelley, K. K., Contributions to the Data on Theoretical Metallurgy. VII. The Thermodynamic Properties of Sulfur and Its Inorganic Compounds. BuMines B 406 (1937).

30. Kelley, K. K., Contributions to the Data on Theoretical Metallurgy. VIII. The Thermodynamic Properties of Metal Carbides and Nitrides. BuMines B 407 (1937).

31. Kelley, K. K., Contributions to the Data on Theoretical Metallurgy. IX. The Entropies of Inorganic Substances. Revision (1940) of Data and Methods of Calculation. BuMines B 434 (1941).

32. Kelley, K. K., Contributions to the Data on Theoretical Metallurgy. X. High-Temperature Heat-Content, Heat-Capacity, and Entropy Data for Inorganic Compounds. BuMines B 476 (1949).

33. Kelley, K. K., Contributions to the Data on Theoretical Metallurgy. XI. Entropies of Inorganic Substances, Rev. (1948) of Data and Methods of Calculation. BuMines B 477 (1950).

34. Coughlin, J. P., Contributions to the Data on Theoretical Metallurgy. XII. Heats and Free Energies of Formation of Inorganic Oxides. BuMines B 542 (1954).

35. Kelley, K. K., Thermodynamic Properties of Zirconium Compounds. Ch. 4 of Metallurgy of Zirconium, ed. by B. Lustman and F. Kerze, Jr. McGraw-Hill Book Co. New York, p. 59 (1955).

36. Kelley, K. K., and A. D. Mah, Metallurgical Thermochemistry of Titanium. BuMines RI 5490 (1959).

37. Kelley, K. K., Contributions to the Data on Theoretical Metallurgy. XIII. High-Temperature Heat-Content, Heat-Capacity, and Entropy Data for the Elements and Inorganic Compounds. BuMines B 584 (1960).

38. Kelley, K. K., and E. G. King, Thermodynamics of Hafnium and Its Compounds Ch. 9 of the Metallurgy of Hafnium, ed. by D. E. Thomas, and E. T. Hayes, Naval Reactor Handbook, p. 323 (1960).

39. Mah, A. D., Thermodynamic Properties of Manganese and Its Compounds. BuMines RI 5600 (1960).

40. Kelley, K. K., and E. G. King, Contributions to the Data on Theoretical Metallurgy. XIV. Entropies of the Elements and Inorganic Compounds. BuMines B 592 (1961).

41. Kelley, K. K., Contributions to the Data on Theoretical Metallurgy. XV. A Reprint of B 383, 384, 393, and 406. BuMines B 601 (1962).

42. Mah, A. D., Thermodynamic Properties of Vanadium and Its Compounds. BuMines RI 6727 (1966).

43. King, E. G., A. D. Mah, and L. B. Pankratz. Thermodynamic Properties of Copper and Its Inorganic Compounds. INCRA Monograph Series II (sponsored by the International Copper Research Association and the U.S. Bureau of Mines), New York, (1973).

44. Mah, A. D., and L. B. Pankratz, Contributions to the Data on Theoretical Metallurgy. XVI. Thermodynamic Properties of Nickel and Its Inorganic Compounds. BuMines B 668 (1976).

45. L. B. Pankratz, Thermodynamic Properties of Elements and Oxides. BuMines B 672, (1982).

46. Carroll W. DeKock, Thermodynamic Properties of Selected Transition Metal Sulfates and Their Hydrates. BuMines IC 8910, (1982).

47. A. D. Mah, Thermodynamic Data for Arsenic Sulfide Reactions. BuMines RI 8671, (1982).

48. R. Hultgren, P. D. Desai, D. T. Hawkins, M. Gleiser, and K. K. Kelley, "Selected Values of the Thermodynamic Properties of Binary Alloys," ASM, Metals Park, Ohio (1973).

49. A. R. Miedema, and P. F. de Chatel, in "Theory of Alloy Phase Formation," edited by L. H. Bennett, Conference Proc., TMS-AIME, p. 334 (1980).

50. R. Boom, F. R. deBoer, and A. R. Miedema, J. Less Common Met., 45, 237 (1976); 46; 271 (1976).

51. P. C. P. Bouten and A. R. Miedema, J. Less-Common Met. 65, 217 (1979); 71, 147 (1980).

52. N. A. Gokcen, "Statistical Thermodynamics of Alloys" Plenum Press (to be published in 1984).

53. O. Kubaschewski, High Temp. High Press. 13, p. 435 (1981).

BIBLIOGRAPHY

B-1 Bennington, K. O., M. J. Ferrante, and J. M. Stuve, Thermodynamic Data on the Amphibole Asbestos Minerals Amosite and Crocidolite. BuMines RI 8265 (1978).

B-2 Stuve, J. M., M. J. Ferrante, and H. C. Ko, Thermodynamic Properties of $NiBr_2$ and $NiSO_4$ From 10-1200 K. BuMines RI 8271 (1978).

B-3 Ko, H. C., M. J. Ferrante, and J. M. Stuve, Thermophysical Properties of Acmite, Proceedings of the 7th Symposium on Thermophysical Properties. Am. Soc. of Mech. Engr, 392 (1978).

B-4 Schaefer, S. C., Electrochemical Determination of the Gibbs Energy of Formation of Sphalerite. BuMines RI 8301 (1978).

B-5 Ferrante, M. J., J. M. Stuve., G. E. Daut, and L. B. Pankratz, Low-Temperature Heat Capacities and High Temperature Enthalpies of Cuprous and Cupric Sulfides. BuMines RI 8305 (1978).

B-6 Ferrante, M. J., D. Cubbicciotti, and K. H. Lau, Thermodynamics of Vaporization and High Temperature Enthalpy of Zirconium Tetraiodide. J. Elec. Soc., v. 125, 972 (1978).

B-7 Beyer, R. P., and H. C. Ko, Low-Temperature Heat Capacities and Enthalpy of Formation of Copper Difluoride. BuMines RI 8329. (1978).

B-8 Gokcen, N. A., Water and magmas: clarification of a controversy on applications of the Gibbs-Duhem equation. Geochim. et Cosmo. Acta, v. 43. (1979).

B-9 Beyer, R. P., and G. E. Daut, Low-Temperature Heat Capacities of Potassium Disilicate. J. Chem. and Eng. Data (1979).

B-10 Gokcen, N. A., and J. J. Loferski, Efficiency of Tandem Solar Cell Systems As A Function of Temperature and Solar Energy Concentration Ratio. Solar Energ. Mat. 1 (1979).

B-11 Gokcen, N. A., Determination and Estimation of Ionic Activities of Metal Salts in Water. BuMines RI 8372 (1979).

B-12 Brown, R. R., G. E. Daut, R. V. Mrazek, and N. A. Gokcen, Solubility and Activity of Aluminum Chloride in Aqueous Hydrochloric Acid Solutions. BuMines RI 8379 (1979).

B-13 Schaefer, S. C., and N. A. Gokcen, Thermodynamic Properties of Liquid Al-Ni and Al-Si Systems. High Temp. Sc., v. 11, 31 (1979).

B-14 Schaefer, S. C., Electrochemical Determination of Gibbs Energies of Formation of MoS_2 and WS_2. BuMines RI 8405 (1980).

B-15 Ko, H. C., and G. E. Daut, Enthalpies of Formation of α- and β-Magnesium Sulfate and Magnesium Sulfate Monohydrate. BuMines RI 8409 (1980).

B-16 Beyer, R. P., M. J. Ferrante, R. R. Brown, and G. E. Daut, Thermodynamic Properties of Potassium Metasilicate and Disilicate. BuMines RI 8410 (1980).

B-17 Ferrante, M. J., and R. A. McCune, High-Temperature Enthalpy and X-Ray Powder Diffraction Data for ZrI_4. BuMines RI 8418 (1980).

B-18 Stuve, J. M., M. J. Ferrante, D. W. Richardson, and R. R. Brown, Thermodynamic Properties of Ferric Oxychloride and Low-Temperature Heat Capacity of Ferric Trichloride. BuMines RI 8420 (1980).

B-19 Stuve, J. M., and M. J. Ferrante, Low-Temperature Heat Capacities and High Temperature Enthalpies of Chiolite. BuMines RI 8442 (1980).

B-20 Bennington, K. O., J. M. Stuve, and M. J. Ferrante, Thermodynamic Properties of Petalite. BuMines RI 8451 (1980).

B-21 Gokcen, N. A., Partial Pressures of Gaseous HCl and H_2O over Aqueous Solutions of HCl, $AlCl_3$, and $FeCl_3$. BuMines RI 8456 (1980).

B-22 Beyer, R. P., M. J. Ferrante, and R. R. Brown, Thermodynamic Properties of $KAlO_2$. J. Chem. Thermo., 985 (1980).

B-23 Gokcen, N. A., Retrograde Solubility in Binary Systems. Scripta Metal., v. 14, 1185 (1980).

B-24 Neumann, J. P., On the Occurrence of Substitutional and Triple Defects in Intermetallic Phases With the B2 Structure. Acta Metal., v. 28, 1165 (1980).

B-25 Schaefer, S. C., Electrochemical Determination of Gibbs Energies of Formation of MnS and $Fe_{0.9}S$. BuMines RI 8486 (1980)

B-26 Schaefer, S. C., and N. A. Gokcen, Electrochemical Determination of Thermodynamic Properties of Molybdenite, High Temp. Sci., v. 12, 267 (1980).

B-27 Ferrante, M. J., and R. A. McCune, High-Temperature Enthalpy and X-Ray Powder Diffraction Data for Aluminum Sulfide, BuMines RI 8526 (1981).

B-28 Beyer, R. P., An Algorithm for Determining Debye Temperatures, BuMines RI 8566 (1981).

B-29 Ferrante, M. J., J. M. Stuve, and L. B. Pankratz, Thermodynamic Properties of Cuprous and Cupric Sulfides, High Temp. Sci., v. 14, 77 (1981).

B-30 Ferrante, M. J., J. M. Stuve, H. C. Ko, and R. R. Brown, Thermodynamic Properties of Aluminum Sulfide, High Temp. Sci., v. 14, 91 (1981).

B-31 Schaefer, S. C., Electrochemical Determination of Gibbs Energies of Formation of Cobalt and Nickel Sulfides, BuMines RI 8588 (1981).

B-32 Schaefer, S. C., and N. A. Gokcen, Thermodynamic Properties of Ferrous and Ferric Fluoride by EMF Measurements. High Temp. Sci., v. 14, 153 (1981).

B-33 Gokcen, N. A., Chemical Metallurgy-A Tribute to Carl Wagner, Proceedings of a symposium sponsored by TMS-AIME Phys. Chem. Comm., Feb 23-25 (1981).

B-34 Schaefer, S. C., and N. A. Gokcen, Electrochemical Determination of Thermodynamic Properties of MnS, Chem Metal - A Tribute to Carl Wagner. Proceedings of a symposium sponsored by TMS-AIME Phys. Chem Comm., at 110th AIME Annual meeting, Chicago, Il, Feb 23-25, 1981, ed. by N. A. Gokcen, 97 (1981).

B-35 Gokcen, N. A., R. V. Mrazek, and L. B. Pankratz, Proceedings of the Workshop on Techniques for Measurement of Thermodynamic Properties. BuMines IC 8853 (1981).

B-36 Beyer, R. P., Automation of a Low-Temperature Calorimeter. BuMines IC 8853, p. 113 (1981).

B-37 Ferrante, M. J., High-Temperature Enthalpy Measurements With a Copper-Block Calorimeter. BuMines IC 8853, p. 134 (1981).

B-38 Stuve, J. M., Sulfide Solution Calorimetry - A Novel Method. BuMines IC 8853, p. 161 (1981).

B-39 Ko, H. C., Determination of Enthalpies of Formation by Solution Calorimetry.

B-40 Bennington, K. O., Some Techniques and Measurements With HF Solution Calorimetry. BuMines IC 8853, p. 173 (1981).

B-41 Schaefer, S. C., Electrochemical Determination of Gibbs Energy of Formation of Sulfides. BuMines IC 8853, p. 203 (1981).

B-42 Neumann, J. P., and N. A. Gokcen, Determination of Ionic Activities of Aqueous Metal Salt Solutions, BuMines IC 8853, p. 214 (1981).

B-43 Mah, A. D., Survey of Bureau of Mines Critical Compilations. BuMines IC 8853, p. 305 (1981).

B-44 Pankratz, L. B., Thermodynamic Data Evaluation. BuMines IC 8853, p. 309 (1981).

B-45 Bennington, K. O., and R. R. Brown, Thermodynamic Properties of Synthetic Acmite. BuMines RI 8621 (1982).

B-46 Ko, H. C., N. Ahmad, and Y. A. Chang, Thermodynamics of Calcination of Calcite. BuMines RI 8657 (1982).

B-47 Gokcen, N. A., Multicomponent Regular Solutions. Scripta Metal., v. 16, 723 (1982).

B-48 Ko, H. C., and R. R. Brown, Enthalpies of Formation of $ZnO \cdot 2ZnSO_4$ and $CoSO_4 \cdot 6H_2O$. BuMines RI 8688 (1982).

B-49 Beyer, R. P., A Computer Program For Calculating Thermodynamic Properties From Spectroscopic Data. BuMines IC 8871 (1982).

B-50 Mah, A. D., Thermodynamic Data for Arsenic Sulfide Reactions. BuMines RI 8671 (1982).

B-51 Mah, A. D., Chemical Equilibria in Chlorination of Clay. BuMines RI 8696 (1982).

B-52 Schaefer, S. C., Electrochemical Determination of Thermodynamic Properties of Manganomanganic Oxide and Manganese Sesquioxide. BuMines RI 8704 (1982).

B-53 Stuve, J. M., A Novel Bromine Calorimetric Determination of the Formation Enthalpies of Sulfides. BuMines RI 8710 (1982).

B-54 Bennington, K. O., Stability Relationships Between Petalite and Spodumene. BuMines RI 8719 (1982).

B-55 Schaefer, S. C., and N. A. Gokcen, Electrochemical Determination of the Thermodynamic Properties of Sphalerite, ZnS (β). High Temp. Sci., v. 15, 225 (1982).

B-56 Beyer, R. P., R. R. Brown, K. O. Bennington, and M. J. Ferrante, Heat Capacity From 5 to 300 K of α- and β-TiO, Relative Enthalpy to 1200 K, Enthalpy of Transition, and Thermodynamic Properties to 1265 K. J. Chem. Thermo., v. 14, 957 (1982).

B-57 Stuve, J. M. Low-Temperature Heat Capacities of Sodium Hexatitanate. J. Chem. and Engr. Data, v. 27,

B-58 DeKock, Carroll W. Thermodynamic Properties of Selected Transition Metal Sulfates and Their Hydrates. BuMines IC 8910 (1982).

B-59 Ferrante, M. J., and N. A. Gokcen. Relative Enthalpies of Ni_3S_2. BuMines RI 8745 (1982).

B-60 Pankratz, L. B., Thermodynamic Properties of Elements and Oxides. BuMines B 672 (1982).

B-61 Gokcen, N. A., Statistical Thermodynamics of Long Range Order. Scripta Metal., v. 17, 53 (1983).

B-62 Ko, H. C., and R. R. Brown, Enthalpy of Formation of $2CdO \cdot CdSO_4$. RI 8751 (1983).

B-63 Gokcen, N. A., Activity Coefficients of Solutes in Binary Solvents. High Temp. Sci., v. 15, 293 (1983).

B-64 Gokcen, N. A., Determination, Estimation, and Correlation of Activities in Hydrometallurgical Ionic Solutions. Publ. in Proceedings of the Hydrometallurgy, 112th AIME Annual Meeting, Atlanta, Georgia, March 6-10, 1983; 329 (1983).

B-65 Bennington, K. O., and R. R. Brown, The Enthalpy of Formation of Synthetic Cancrinite, RI 8778 (1983).

B-66 Bennington, K. O., R. P. Beyer, and G. Johnson, Thermodynamic Properties of Pollucite (a Cesium-Aluminum-Silicate),

RI 8779 (1983).

B-67 Beyer, R. P. Heat Capacity of TiS_2 From 4.97 to 300.7 K. J. Chem. and Engr. Data (1983).

B-68 Schaefer, S. C. Electrochemical Determination of the Thermodynamic Properties of Manganese Sulfate and Cadmium Oxysulfate. BuMines RI 8809 (1983).

B-69 Gokcen, N. A. Rates of Chlorination of Aluminous Resources. BuMines IC 8952, (1983).

B-70 Beyer, R. P., M. J. Ferrante, and R. V. Mrazek. An Automated Calorimeter for Heat Capacity Measurement From 5 to 300 K. The Heat Capacity of Cadmium Sulfide From 4.8 to 301.8 K, and the Relative Enthalpy to 1200 K. J. Chem. Thermo., 15 (1983).

B-71 Beyer, R. P. Heat Capacities of Zinc Oxysulfate From 5.1 to 309.4 K With Transition at 279 K. J. Chem. Thermo., 15 (1983).

B-72 Brittain, S. L. Thermodynamic Research 1915-1983. A list of thermodynamic research publications. Albany Research Center, Albany, OR (1983).

SOURCES OF THERMOCHEMICAL DATA: ARCHIVAL AND CURRENT

Robert D. Freeman ■ Chemistry Department, Oklahoma State University, Stillwater, OK 74078

The question addressed is not "Where does one find data for property X on compound Z?", but "Where does one find *information about publications* (both archival and current) in which thermochemical data of various types may be found?"

ARCHIVAL

The chemical engineer frequently needs to find 'in the literature' values for various thermodynamic properties. Unless s/he happens to be expert in the area, the reasonable first question, upon encountering such need, is: 'Is there a guide to compilations of thermodynamic data which will tell me where I'm most likely to find what I need?'. Fortunately, the answer is now 'yes'.

The International Council of Scientific Unions (ICSU) established some years ago a Committee on Data (CODATA), charged with promoting the production and dissemination of scientific and technical data. One of CODATA's projects is the *CODATA Directory of Data Sources for Science and Technology*. At some point this Directory will be published in book form; in the meantime each individual chapter is being published, as it is completed, as a 'CODATA Bulletin'. Chapters published through 1982, with the identifying 'Bulletin' number in parentheses, are: Crystallography (24), Hydrology (35), Astronomy (36), Zoology (38), Seismology (42), Chemical Kinetics (43), Nuclear and Elementary Particle Physics (48), Atomic and Molecular Spectroscopy (49), and Geodesy. CODATA Bulletin Nos. 1 – 39 are available from the CODATA Secretariat(*1*); those numbered 40 and onwards have been published by Pergamon(*2*).

A chapter on 'Chemical Thermodynamics' for the *CODATA Directory* has been in progress for several years and is now essentially complete; publication is expected in early 1984(*3*). For convenience, this chapter is hereinafter referred to as the *CT Directory*.

CT Directory represents a first attempt (there will undoubtedly be revisions and updates) to provide comprehensive detailed information on 'Data Centers' and their output. A 'Data Center' is defined as a group of individuals who have been and/or who reasonably expect to be compiling, evaluating, and publishing chemical thermodynamic data *over an extended time*. The *CT Directory* contains information on some 60 Data Centers, about half in the USA and UK, the other half distributed world-wide. For each center, information is given specifically on the substances for which data are provided and on the thermodynamic properties covered. References to 'output' are given, usually books or journal articles.

In addition, the *CT Directory* contains a rather *large, annotated bibliography* of sources of thermodynamics data – about 200 direct entries and another 150 or so indirect entries. With few exceptions, each entry is for a non-journal publication which contains a compilation of data, usually critically evaluated. The entries are arranged under 12 headings, e.g., 'Fluids: Pure Gases and Liquids', 'Mixtures: Vapor-Liquid Equilibria', with 5 – 25 entries under each heading. Finding and scanning the entries pertinent to a specific need can be done rather quickly.

It should be noted, perhaps, that the orientation of the *CT Directory* is 'chemical', not 'chemical engineering', thermodynamics. Nevertheless, there is very much overlap between the two subdisciplines, and the chemical engineer who wants to find compilations of thermodynamic data should find the *CT Directory* to be *the* place to start. There are, of course, other bibliographies of thermodynamic data. They are not listed here because there has been a major effort to include in the *CT Directory* all significant sources listed in any other known bibliography of thermodynamic data.

CURRENT

Finding an authorative compilation of the desired type of data does not necessarily solve one's problem. Sometimes data for the substance of interest are not in that compilation — nor in any other. The only recourse is the recent literature, or work *not yet* in the literature — if one can find out about it. There is, of course, *Chemical Abstracts*, but a manual search can be tedious and time-consuming and an automated search by keyword requires the user and the indexer to 'think alike' in choosing key words. Fortunately, for chemical thermodynamic data there is a solution: the annual *Bulletin of Chemical Thermodynamics*(4).

The *Bulletin* contains three major sections: Index, Bibliography, and Reports, which are described in more detail below. Briefly, the Bibliography provides a listing of papers with chemical thermodynamic content published during the preceeding year; the Reports provide brief summaries of work completed but not yet published; the Index provides rapid searching of and entry to the Bibliography and the Reports in terms of the Chemical Substance – Thermodynamic Property discussed or described in a bibliographic citation or in a Report.

Reports

The Reports Section provides terse summaries [i.e., what substance(s), what properties or information, over what range of variables] of research/measurements *completed but not yet published*, in some 500 labs from Lund, Sweden to Bundoora, Australia, from Gorky, USSR to Berkeley, USA. The reports, submitted by scientists from some 35 countries, are grouped into Sections — labeled I, H, K, M, P, Q, V, X, Y, Z — according to the research described in the Report. The following descriptions indicate the types of studies/information included in each Section.

I: Indentification of, and brief general statement about the chemical thermodynamic research interests of, contributing investigators/institutions.

H: Enthalpy Changes for Process and Reactions by Calorimetry: combustion, reaction, mixing, adsorption, etc.

K: Reaction Equilibria and Related Data: equilibrium constants for reactions, dissociation/decomposition pressures, electrochemical cell potentials, and derived quantities.

M: Thermodynamic Quantities Calculated from Molecular Parameters: tabulation of functions for ideal gases, intermolecular potentials/derived equations of state for real gases, bond energies, etc.

P: Phase Equilibria: transition temperatures, vapor pressure, vapor/liquid equilibria, condensed phase equilibria (solubilities, phase diagrams, etc.), surface phenomena, etc., and derived quantities.

Q: Thermal Properties for Non-Reacting Systems by Calorimetry: heat capacity, enthalpy, entropy, Planck function, Gibbs function, etc., including phase transitions.

V: Volume as $f(T,p,x)$; Empirical Equations of State: $pVT(x)$ data, density, critical state properties, compressibility, thermal expansivity, excess volume, etc.

X: Physical Properties of a Single Phase: surface tension, viscosity, refractive index, etc.

Y: Biochemical and Macromolecular Systems: chemical thermodynamic quantities, however determined, for biochemical and/or macromolecular substances.

Z: Compilations and Correlations: literature searches/bibliographies, critically evaluated compilations, reviews, correlations of thermal and thermochemical properties, etc.

Index

The typical user of the Bulletin is interested in finding information on some particular thermodynamic property for a specific chemical compound, or group of compounds. The Index is organized specifically to assist the user in that search: it is a Chemical Substance – Thermodynamic Property Index.

Each chemical substance (element, compound, mixture, alloy, solution, etc.) discussed in an article in the Bibliography or in a Report is listed in the Index in a standardized, rational arrangement, the reported property is indicated in the Index by a simple mnemonic code, and a code number indicates where to find the article in the Bibliography or in the Reports Section.

Bibliography

The Bibliography is a listing of author(s), title, and source of articles with chemical thermodynamic content – very broadly interpreted – published during the preceeding calendar year. In a typical volume, there are some 1300 references in the Organic Section, 1600 in the Organic Systems (Mixtures) Section, 2400 in Inorganic, and 300 in Biochemical and Macromolecular Systems — a total of about 5600 references to thermodynamic data/information.

The Bibliography is generated by four groups of scientists, each responsible for a specific Section: Organic Substances Section by the Thermodynamics Research Center/Texas A&M University; Organic Systems (Mixtures) Section by the Institute for Topology and Dynamics of Systems/University of Paris VII; Inorganic Substances Section by the Chemical Thermodynamics Data Center/National Bureau of Standards; Biochemical and Macromolecular Systems Section by the Thermochemical Institute/Brigham Young University.

These groups search some 60 scientific and engineering journals, article by article, for papers with chemical thermodynamic content. In addition, various abstracting publications, including the pertinent sections of Chemical Abstracts, are scanned for relevant material.

Other regular features of the *Bulletin* include: calendar of meetings, book reviews, a listing of recently published books with thermodynamic content, Reports from the Calorimetry Conference, Reports from IUPAC and CODATA Commissions/Task Groups, and other miscellaneous items. Review papers on various thermodynamic topics appear occassionally.

In short, the *Bulletin* is, and for 25 years has been, an annual comprehensive summary of worldwide activity in chemical thermodynamics – with 'chemical thermodynamics' interpreted very broadly: from heats of combustion to non-equilibrium statistical thermodynamics; from equilibrium constants and emfs of cells to $pVT(x)$ data, vapor-liquid equilibria, and property correllations. Coverage includes all non-transport thermodynamic properties of chemical substances.

Work is currently in progress to make the Index and Bibliography searchable by computer. Much remains to be done but there is a reasonable expectation that automated searching will be available within a couple of years.

ON-LINE DATA SYSTEMS

There is a steadily growing trend to 'put everything

on the computer'. Thermodynamic data are prime candidates, and numerous 'databases' have been established; many are private, and some are available, on-line, for public use. The *CT Directory*(3)(see above) contains a section which identifies and describes on-line thermodynamic data systems (or databases) available to more-or-less anyone. Sufficient detail is given for one to establish a probability of serious interest; name, address, and phone number are given for the source of more information.

ADJUVANT DATA

Thermodynamic analysis of a particular system or process often requires a variety of data *other* than the typical 'thermochemical' data discussed above. Obvious examples of needed data are molecular structural parameters, atomic energy levels and ionization energies, and bond energies.

An attempt has been made, in cooperation with the NBS Chemical Thermodynamics Data Center, the JANAF Thermochemical Tables Group, and the Texas A&M Thermodynamics Research Center, to compile a bibliography of such adjuvant data sources. The first attempt appeared in 1979(5); a revised version is to be published (5) and occassional updates are expected. A second very useful compilation is one by Westbrook and Desai(6).

SUMMARY

The two key publications for locating sources of thermodynamic data are: the 'Chemical Thermodynamics' Chapter(3) of the *CODATA Directory*, for currently active Data Centers, for archival compilations, and for on-line databases; and the *Bulletin of Chemical Thermodynamics*(4) for broad, comprehensive coverage of current literature and of work completed but not yet in the literature. Bibliographies of adjuvant data are available(5,6).

LITERATURE CITED

1. CODATA Secretariat, 51 Boulevard de Montmorency, F-75016 Paris, France.
2. Pergamon Press: Headington Hill Hall, Oxford OX3 OBW, UK; Maxwell House, Fairview Park, Elmsford, NY 10523 USA.
3. R.D. Freeman, Chapter 11: 'Chemical Thermodynamics', in *CODATA Directory of Data Sources for Science and Technology*, CODATA Bulletin No. 55, 1984. Pergamon Press, Oxford.
4. R.D. Freeman, ed. *Bulletin of Chemical Thermodynamics,* **23**(1980), **24**(1981), **25**(1982). Thermochemistry, Inc., Oklahoma State University, Stillwater, OK 74078 USA.
5. R.D. Freeman, *Bull. Chem. Thermodyn.* **22**, 499–505(1979); *ibid.*, **26**(1983).
6. J.H. Westbrook, J.D. Desai, 'Data Sources for Materials Scientists and Engineers', in *Ann. Rev. Materials Science* **8**, 359–422 (1978).

AQUEOUS SOLUTIONS PROPERTIES AT HIGH TEMPERATURES AND PRESSURES

Sidney L. Phillips and Lenard F. Silvester ■ Lawrence Berkeley Lab., Univ. of California, Berkeley, CA 94720

A correlation equation is given which reproduces the density of sodium chloride solutions to ±4% over the ranges 0-5 molal, 0-350°C and 1-1000 bars. Data generated from the equation are compared with selected experimental and smoothed values at typical temperatures, pressures and concentrations. The equation is applied to calculation of the ion product of water up to 350° and 1000 bars.

Gaps in information are the unfortunate consequences of the diligence with which scientists and engineers devote themselves to immediate problems. However, delays in transferring new data from research to the user thereby reducing gaps are minimizable. Transfer is made quickly and efficiently by specialized databases which compile critically evaluated data using modern computer assisted methods for data handling.

Thermochemical values for estimating chemical equilibria to high temperatures; and, thermophysical data on the properties of substances such as water are effectively handled with correlating equations. These equations are theoretically based, or are the result of empirical fits. Both are important; however, theoretically based equations are extrapolated with more confidence.

The development of new energy resources and advanced chemical processes requires sound thermophysical data on aqueous solutions to high temperatures. The combined need is for modern data in a quickly useful form. This paper gives a correlating equation used to generate tables of data on the density of sodium chloride solutions up to 350° C, 1-1000 bars and 0-5 molal concentrations. Such data are used to predict

Lawrence Berkeley Laboratory, University of California, Berkeley, CA 94720

characteristics of brines for geothermal energy (1-3), sea water desalination and for the origin of subsurface brines. The equations were developed by fitting critically evaluated experimental data (4).

The critical evaluation and selection of experimental data were based on the following criteria: details given on the experimental procedure; purity of materials; uncertainty assigned by the investigator; number of replicate measurements; temperature, pressure and concentration range covered; publication in refereed journals; and, prior publications by the researcher. Interpolated values from these correlating equations were then compared to both experimental and calculated data.

Density of Sodium Chloride Solutions

Experimental data on the density of sodium chloride solutions were selected and fit to the following empirical equations:

$$d = A + Bx + Cx^2 + Dx^3 \tag{1}$$

$$x = c_1 \exp(a_1 m) + c_2 \exp(a_2 T) + c_3 \exp(a_3 P) \tag{2}$$

d = density, g/cm³; m = molality; T = °C; P = bars
A = -3.033405; B = 10.128163;
C = -8.750567; D = 2.663107
c_1 = -9.9559; c_2 = 7.0845; c_3 = 3.9093

$a_1 = -4.539\,E-003$; $a_2 = -1.638\,E-004$; $a_3 = 2.551\,E-005$

$0 < T < 350°\,C$; $1 < P < 1000\ bars$

$0 < m < 5\ molal$

Some features of eq 1 are the following: calculation of temperature and pressure derivatives is straightforward; the equation is easily programmed; and experimental data are reproduced with sufficient accuracy for many applications. Equation 1 interpolates the density of water to $\pm$ 4% deviation for m=0, 1-1000 bars, and 0-350° C, when compared to 79 values selected from the recent publications by Rogers, Bradley and Pitzer; Out and Los; Nagashima; and from the Steam Tables. Comparison of data for the density of water from eq 1 when m=0 in eq 2 with values calculated from Rowe and Chou for 0-100° C, and from Isdale and Morris for 70-160° C were also $\pm$ 4% difference.

Similarly, the density of NaCl solutions is reproduced to $\pm$ 4% by eq 1, when compared to 200 points selected from the following: Out and Los; Rogers, Bradley and Pitzer; Grant-Taylor; Zarembo and Fedorov; Gorbachev, Kondrat'ev, Androsov and Kolupaev; and Ellis. We do not include data from Zarembo et al. for 350° C at 24% (5.404m) concentration because of the comparatively larger error (-5.7% to -7.8%). Similarly, the last six values for 1.000 m solutions for Table 1 of Grant-Taylor are erroneous and, therefore, not included here. The density values published by Khaibullin and Novikov for 100-417° C, 1-338 kg/cm^2, 1% and 5% of NaCl (0.1728 and 0.9006 m) gave the largest errors (up to 15%), and consequently are not included in these comparisons.

A plot of % deviation in density for 112 points over the pressure range 20-1000 bars using data from Rogers et al.; Zarembo and Fedorov; Gorbachev et al.; and Grant-Taylor show that above 500 bars, values interpolated by eq 1 are 0-3% higher than e.g., Zarembo and Fedorov. A similar plot shows % deviation for 107 points for 0.1-4 m concentrations in comparing eq 1 with the recent publications by Out and Los; Rogers et al. and Grant-Taylor, that there is a noticeable trend of zero to about 3% lower results as the concentration increases from 2-4 m. A comparable plot of 117 points of data from Ellis; Gorbachev et al.; and, Zarembo and Fedorov indicates a similar trend.

Our values for NaCl solutions were compared with specific volumes calculated by the equation for volumetric properties published recently by Rogers and Pitzer covering temperatures to 300° C and pressures to 1000 bars. Agreement is quite good generally, with the largest difference of about $\pm$ 1.5% apparently at 1000 bars. A comparison of the two sets of data as a function of concentration with the pressure fixed at 200 bars, and temperatures of 100° C and 300° C was made. At 100° C, the two sets begin to diverge near 1 molal NaCl to a maximum difference of 2.6% at 5 m. For 300° C, the two curves merge at about 2m NaCl, with maximum differences of 2.9% at 0.1 m, and 0.8% at 5 m. A comparison for 0-5 m, 1 bar and 100° C gives two curves with a difference of 0.5% at 0.1 m, which increases to 3% at 5 m. The agreement is considered satisfactory, given the different databases used to develop the correlations.

Comparison of our data for NaCl solutions with selected densities from Rowe and Chou indicates at any constant pressure and molal concentration, % deviation is usually positive for temperatures from 0-150° C.

Table 1 lists selected sources of experimental data on density of NaCl solutions.

Ion Product of Water

Interpolating equations for the ion product and dielectric constant of water utilize the density of water. For example, the ion product of water, $H_2O = H^+ + OH^-$, is reproduced by the following equation recently published by Marshall and Franck.

$$\log K_w = -4.098 - \frac{3245.2}{T} + \frac{2.2362\,E+005}{T^2} - \frac{3.984\,E+007}{T^3} \quad (3)$$

$$+ \left(13.957 - \frac{1262.3}{T} + \frac{8.5641\,E+005}{T^2}\right)\log d_w$$

where $T = 273.15 + °C$

Note that in eq 3 the pressure dependence for log K_w is contained in the density term. We have calculated values of log K_w using eq 1 for water density, at representative pressures of 1, 500 and 1000 bars. Figure 1 shows agreement between the results from eq 1 and eq 3 with those tabulated by Marshall and Franck. The percentage deviation varies with temperature, and is within ± 3%. See Figure 2. Equation 1 for density was developed for temperatures up to 350° C, so that log K_w calculated with these equations is probably valid only up to 350° C and 1000 bars (4). Because more accurate correlations are available up to 150° C, eq 1 is most useful for temperatures above 150° C. See Table 2.

Units and Conversions

1 bar = 1.019716 kg/cm^2

1 atm = 1.01325 bar = 1.03323 kg/cm^2

% weight NaCl = (100)(molality)x (58.44)/(1000 + (molality)(58.44))

REFERENCES

(1) Phillips, S.L.; Igbene, A.; Fair, J.A.; Ozbek, H.; Tavana, M.; "A Technical Databook for Geothermal Energy Utilization," Lawrence Berkeley Laboratory: Berkeley, CA 94720; LBL-12810; June 1981.

(2) Phillips, S.L.; Ozbek, H.; Igbene, A.; Litton, G.; "Viscosity of NaCl and Other Solutions Up to 350° C and 50 MPa Pressures," Ibid.; LBL-11586; November 1980.

(3) Ozbek, H.; Phillips, S.L.; J. Chem. Eng. Data 1980, 25, 263.

(4) Phillips, S.L.; Ozbek, H.; Silvester, L.F.; Lawrence Berkeley Laboratory: Berkeley, Ca 94720; LBL-16275; June 1983. High Temperatures-High Pressures 1983, V.15.

Acknowledgement

Thanks are given to C. Bethke, University of Illinois; D.F. Grant-Taylor, Department of Scientific and Industrial Research; and, P.S.Z. Rogers, Los Alamos National Laboratory for their comments.

This Work was supported by the Director, Office of Energy Research, Office of Basic Energy Sciences, Engineering & Geosciences Division of the U.S. Department of Energy under Contract No. DE-AC0376SF00098.

Table 1. Selected experimental data for density of water and sodium chloride solutions to high temperatures and pressures. Some data are published in the form of a correlation equation.

Ranges	Measurement	Reference
25-45;1;1-5	Oscillating tube	Romankiw;Chou(1983)
5-34;1;0.37802-5.99684	Vibrating tube	LoSurdo;Alzola;Millero(1982)
30-200;20.3;0-4.39	Dilatometer	Rogers;Bradley;Pitzer(1982)
173-354;200;0.100-4.000	Mercury displacement	Grant-Taylor(1981)
0-100;1;0	Calculated	Kell(1981)
5-95;1;0-1.2000	Calculated	Out;Los(1980)
0-35;1;0.00992-1.49986	Vibrating flow densimeter	Chen;Chen;Millero(1980)
25-50;1;0.00784-5.8267	Pycnometer	Goncalves;Kestin(1977)
25-350;0-980.7;0.35-5.4	Hydrostatic weighing	Zarembo;Fedorov(1976)
0-100;0-1000;0	Calculated	Kell;Whalley(1975)
40-280;100.1;0.001-1.5		Gorbachev;Kondrat'ev;Androsov;Kolupaev(1974)
70-160;SVP;0	Dilatometer	Isdale;Morris(1972)
0-175;1-303;0-5.7	Compressibility	Rowe;Chou(1972)
25-200;20.3;0.1-1	Mercury displacement	Ellis(1966)
25-175;SVP;0.1-2.5	Pycnometer;dilatometer	Fabuss;Korosi;Huq(1966)
0-40;0-1000;0-1	Equation of state	Chen;Millero(1981)

Aqueous Solutions Database — June 1983 NaCl Solutions

Table 2. Density of NaCl solutions g/cm 3, at molal concentration shown

TEMP (C)	PRES BARS	0	.5	1	2	3	4	5
150	10	.90555	.92632	.94538	.97912	1.00828	1.03434	1.05874
150	100	.91403	.93412	.95256	.98530	1.01374	1.03938	1.06363
150	200	.92315	.94251	.96030	.99189	1.01970	1.04492	1.06908
150	500	.94868	.96605	.98210	1.01103	1.03699	1.06141	1.08572
150	1000	.98601	1.00076	1.01462	1.04041	1.06486	1.08940	1.11542
200	100	.85565	.88054	.90332	.94333	.97722	1.00649	1.03258
200	200	.86696	.89091	.91284	.95139	.98415	1.01260	1.03821
200	500	.89854	.91989	.93946	.97405	1.00383	1.03027	1.05482
200	1000	.94419	.96190	.97825	1.00764	1.03387	1.05839	1.08262
250	100	.78392	.81470	.84291	.89240	.93393	.96903	.99918
250	200	.79790	.82754	.85470	.90232	.94232	.97621	1.00547
250	500	.83699	.86343	.88762	.93007	.96588	.99656	1.02357
250	1000	.89346	.91522	.93517	.97038	1.00062	1.02736	1.05204
300	100	.69641	.73416	.76886	.82996	.88127	.92435	.96071
300	200	.71353	.74994	.78339	.84223	.89161	.93309	.96816
300	500	.76157	.79416	.82405	.87651	.92051	.95759	.98923
300	1000	.83127	.85818	.88280	.92601	.96242	.99355	1.02087
350	200	.61155	.65576	.69655	.76869	.82955	.88072	.92370
350	500	.66992	.70972	.74635	.81092	.86522	.91080	.94918
350	1000	.75522	.78832	.81868	.87199	.91670	.95435	.98643

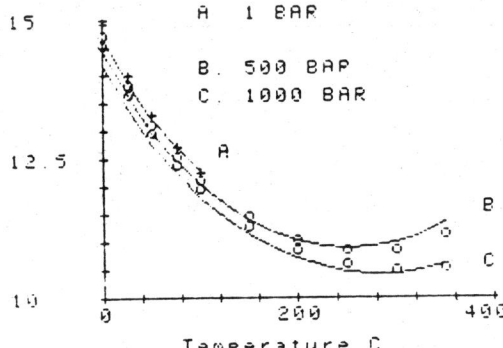

Figure 1. Ion product of water ($-\log K_w$) calculated from eq 3, using densities calculated by eq 1, eq 2 (solid lines). Values are compared with data from Marshall and Franck.

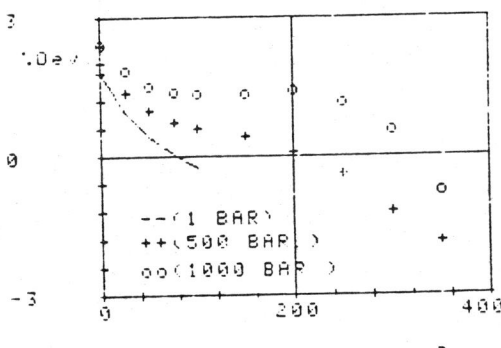

Figure 2. Percentage deviation between ion product of water calculated from eq 3, using densities calculated by eq 1, eq 2 and data from Marshall and Franck, at pressures indicated.

THERMODYNAMIC DATA FOR AQUEOUS SOLUTIONS AND THEIR USES

Loren G. Hepler ■ Department of Chemistry and Department of Chemical Engineering
University of Alberta, Edmonton, Alberta, Canada T6G 2G2

Thermodynamic data for aqueous solutions are useful in many fields of applied chemistry, chemical engineering, geology, mineral processing, etc. The text of this review is concerned with providing a brief summary of some methods of thermodynamics that are applicable to calculations of equalibrium properties of systems involving aqueous solutions along with references to detailed treatments of both principles and applications. Many references to sources of thermodynamic data for aqueous solutions and related systems are also given.

Thermodynamic data for aqueous solutions have many and varied uses in many fields of engineering and applied science, as illustrated by the following incomplete list:

   (i) Boiler water chemistry
  (ii) Coal processing
 (iii) Enhanced oil recovery
  (iv) Geochemical prospecting
   (v) Geothermal energy utilization
  (vi) Heavy oil production
 (vii) Hydrometallurgy
(viii) Radioactive waste disposal
  (ix) Removal of pollutants (H_2S, SO_2, phenol, etc.) from both aqueous and gaseous effluent streams
   (x) Sea (also brackish) water chemistry and water desalination.

Some of these applications have been illustrated recently in an excellent book (1), which is highly recommended.

It is probable that most of the uses of tabulated thermodynamic data for aqueous solutions by chemical engineers and applied chemists (also geologists, mineral processors, etc.) involve calculations of various equilibrium properties, such as the following

---

University of Alberta, Edmonton, Alberta, Canada, T6G 2G2

(all for various conditions of temperature, pressure, and concentrations of solutes):

   (i) Solubilities
  (ii) Reversible electrode potentials
 (iii) pH of solutions containing weak acids, weak bases, hydrolyzable ions or buffers
  (iv) Freezing points
   (v) Vapor pressures
  (vi) Equilibrium extent of various chemical reactions (oxidation-reduction, acid ionization, solute decomposition, etc.).

It is, however, worth noting that tabulated thermodynamic data for aqueous solutions are also useful for other purposes, such as calculations for heat transfer problems that require knowledge of heat capacities (specific heats) and heats of chemical reactions in solutions.

The principal aims of this presentation are as follows:

   (i) To provide references to sources of information about the thermodynamic principles and methods that are pertinent to uses of thermodynamic data for aqueous solutions
  (ii) To provide one general illustration of an application of tabulated thermodynamic data

(iii) To provide an extensive (but not complete) listing of sources of thermodynamic data for aqueous solutions.

THERMODYNAMIC METHODS AND PRINCIPLES

The book (1) titled "Thermodynamics of Aqueous Systems with Industrial Applications" (edited by Newman) contains a substantial number of illustrations of applications of thermodynamics to practical problems involving aqueous solutions. Useful background on the thermodynamic principles and methods used in these and other applications is given in several texts (2-7).

Although the principles and general methods of solution thermodynamics (along with specifics for certain problems) are presented both clearly and correctly in the textbooks (2-7) cited above, not one of these books deals explicitly with one class of problems that is of considerable importance now - namely the use of 25°C data for calculation of equilibrium properties for hot water solutions. I have therefore chosen to describe this kind of calculation in the next section.

A GENERAL ILLUSTRATION

Suppose that we want to use tabulated thermodynamic data in a calculation that will tell us the value of some equilibrium property (such as a solubility) over a range of conditions of temperature. The first step in this calculation is to obtain the equilibrium constant for the chemical reaction of interest at the temperature (often not 25°C) of interest.

In order to be semi-specific while retaining some generality, let us first consider the solubility of a solid salt (such as $BaSO_4$ or $AgCl$) in water as represented by

$$MX(c) = M^{n+}(aq) + X^{n-}(aq) \quad (1)$$

in which (c) indicates a crystalline substance and (aq) indicates a dissolved species. We might also consider the extent of some reaction involving a solid and some aqueous species in solution to yield various other aqueous species. A semi-generalized reaction equation is

$$MX(c) + H^+(aq) = M^{n+}(aq) + HX^{+1-n}(aq) \quad (2)$$

A real example of the kind represented by Equation (2) is the reaction of solid calcium carbonate with dilute aqueous acid to yield dissolved calcium ions and bicarbonate ions.

The most readily available tables of thermodynamic data provide such data for a temperature of 298.15 K (25.00°C). Included in such tables will be the standard state free energies (Gibbs energies) of formation (ΔG_f^o) of most inorganics and "small molecule" organics of interest. Our procedure begins with calculation of the ΔG^o for the reaction of interest, according to the general equation

$$\Delta G^o = \Sigma \Delta G_f^o(\text{products}) - \Sigma \Delta G_f^o(\text{reactants}) \quad (3)$$

in which "products" and "reactants" refer to substances on the right hand and left hand sides, respectively, of the reaction equation. We then use the ΔG^o of reaction calculated according to Equation (3) to obtain the desired equilibrium constant K by way of the equations

$$\Delta G^o = -RT\ln K \quad \text{and} \quad \ln K = -\Delta G^o/RT \quad (4)$$

The equilibrium constant obtained by way of the above calculation naturally refers to the temperature of the tabulated thermodynamic data used in the calculation, which is almost always 298.15 K (25.00°C).

Our next task is to use the K_{298} obtained as just described with other thermodynamic data in a calculation that will yield a value for the same equilibrium constant at some new temperature T, with this new equilibrium constant represented by K_T. This calculation will be based on the van't Hoff equation:

$$(\partial \ln K/\partial T)_p = \Delta H^o/RT^2 \quad (5)$$

To use this Equation (5) we must have information about the enthalpy change (heat) associated with the reaction of interest. The desired ΔH^o of reaction can be calculated from tabulated standard enthalpies (heats) of formation (ΔH_f^o) by way of

$$\Delta H^o = \Sigma \Delta H_f^o(\text{products}) - \Sigma \Delta H_f^o(\text{reactants}) \quad (6)$$

This calculated ΔH^o also refers specifically to the temperature of the tabulated ΔH_f^o values, which is almost always 298.15 K.

The first approximation we make is to take ΔH^o of reaction to be a constant, independent of temperature, which permits integration of Equation (5) to yield

$$\ln K_T = I - \Delta H^o/RT \quad (7)$$

in which I is a constant that can be evaluated from knowledge of ΔH^O and the value of K_{298} at $T = 298.15$. Equation (7) may also be written as

$$\ln K_T = \ln K_{298} + [\Delta H^O/R][(T - 298)/298T] \quad (8)$$

According to Equation (7), $\ln K_T$ should vary linearly with $1/T$. In a general way, $\ln K_T$ does vary linearly with $1/T$ over short ranges of temperature, but over wide ranges of temperature $\ln K_T$ against $1/T$ generally deviates substantially from linearity, with such deviations often leading to a maximum or minimum that is entirely inconsistent with Equations (7) and (8). The result of all this is that Equations (7) and (8) are likely to be useful only for temperatures "near" $T = 298$ K.

The difficulty with the treatment outlined in the preceding paragraph is that ΔH^O is generally dependent on the temperature as indicated by the Kirchhoff Equation (9)

$$(\partial \Delta H^O/\partial T)_p = \Delta C_p^O \quad (9)$$

in which

$$\Delta C_p^O = \Sigma C_p^O(\text{products}) - \Sigma C_p^O(\text{reactants}) \quad (10)$$

Now integration of Equation (9) leads to

$$\Delta H_T^O = \Delta H_{298}^O + \int_{298}^{T} \Delta C_p^O dT \quad (11)$$

as a general expression for ΔH_T^O. Equation (11) shows that it is proper to regard ΔH_T^O as a constant (independent of temperature) only in the special cases of small temperature range or $\Delta C_p^O \cong 0$. We often have $\Delta C_p^O \neq 0$ and are often interested in some temperature T that is not near $T = 298$ K. It is therefore necessary that we have available a treatment that is better than the first approximation that led to Equations (7) and (8).

To remove the often-inadequate approximation that ΔH^O is independent of temperature ($\Delta C_p^O = 0$), we use Equation (11) in Equation (5) to obtain (after some mathematical manipulation)

$$\ln K_T = \ln K_{298} + (\Delta H_{298}^O/298RT)(T - 298)$$

$$- (1/RT) \int_{298}^{T} \Delta C_p^O dT$$

$$+ (1/R) \int_{298}^{T} (\Delta C_p^O/T) dT \quad (12)$$

Use of Equation (12) requires that we know or obtain K_{298} and ΔH_{298}^O as already described earlier in this section. Next we should have ΔC_p^O values over the entire temperature range of interest. We might then fit some algebraic expression (possibly a polynomial) to our ΔC_p^O values and evaluate the integrals in Equation (12) analytically, or we might evaluate these integrals graphically from plots of ΔC_p^O against T and $\Delta C_p^O/T$ against T.

In general, heat capacities of most solids such as MX(c) in Equations (1) and (2) are available or can be estimated satisfactorily. Although our examples have not been concerned with gases in equilibrium with aqueous solutions, such problems are important and can be treated in similar ways. In general, the heat capacities of these gases are also available or can be estimated. Unfortunately for users of thermodynamic calculations, many of the needed heat capacities [partial molar heat capacities (2-4)] of aqueous solutes have not been available in the past. In recent years, however, calorimetric measurements in several laboratories have led to a substantial number of the "missing" partial molar heat capacities of aqueous solutes at 25°C and to enough such heat capacities at other temperatures that a general picture of the temperature dependence of these heat capacities is emerging.

As a result of work mentioned in the preceding paragraph, we now know that it is sometimes an adequate approximation to consider ΔC_p^O to be a constant that is independent of temperature. Using this approximation in Equation (11) leads to

$$\Delta H_T^O = \Delta H_{298}^O + \Delta C_p^O(T - 298) \quad (13)$$

Further combination of Equation (13) with Equation (5) or appropriate evaluation of the integrals in Equation (12) leads to

$$\ln K_T = \ln K_{298} + (\Delta H_{298}^O/298RT)(T - 298)$$

$$- (\Delta C_p^O/RT)(T - 298) + (\Delta C_p^O/R)\ln(T/298) \quad (14)$$

Now that we have the general equation (12) and the useful approximate Equation (14) to use for obtaining $\ln K_T$, it is appropriate to consider the next step in our calculations, which will involve use of K_T. Most readers will remember more or less clearly using equilibrium constants for the solubility of silver chloride and the ionization of acetic acid. For solutions that are reasonably dilute, it is usually a satisfactory approximation to ignore activity coefficients and thence consider only the concentrations of various solute species. However, for more concentrated solutions it is necessary to consider activity coefficients of solute species. These activity coefficients have been discussed thoroughly in books (1-4,7) already cited and also in some other publications that are cited later. Many of the activity coefficients that we are likely to need are already available for solutions at 25°C or can be estimated satisfactorily. Activity coefficients for some solutions at other temperatures have been evaluated by way of appropriate measurements at these temperatures. These activity coefficients for temperatures other than 25°C can also be calculated from activity coefficients at 25°C and appropriate heats of dilution and heat capacities, as described in books (3,4) already cited and in other publications to be cited later.

It may now be clear that thermodynamic data for solutions at temperatures other than 25°C can be obtained relatively directly by way of appropriate measurements at these other temperatures, and can also be obtained less directly but equally accurately by way of appropriate thermodynamic measurements at other (possibly more convenient) temperatures. This last statement is consistent with McGlashan's (5) statement (about thermodynamic equations) that "They are useful because some quantities are easier to measure than others." I now add to these statements the further statement that some thermodynamic quantities are especially useful because they can be used over and over again in a large number of different thermodynamic calculations. Many of these useful thermodynamic quantities for aqueous solutions are already available, ready for use. Others remain to be measured. It is my opinion that the most generally useful quantities for aqueous solutions that are not generally available now are those quantities that are derived from heats of dilution and heat capacities of solutions.

The discussion so far has focused entirely upon the dependence of equilibrium constants on temperature. Exactly the same principles and very similar methods of calculation can be applied to calculation of reversible electrode potentials as a function of temperature.

Effects of pressure on chemical equilibria involving aqueous solutions (and other systems) can also be calculated by appropriate applications of thermodynamic equations. In this case the most convenient starting equation is

$$(\partial \ln K/\partial p)_T = -\Delta V^o/RT \quad (15)$$

in which

$$\Delta V^o = \Sigma V^o(\text{products}) - \Sigma V^o(\text{reactants}) \quad (16)$$

Procedures for doing such calculations have been described by Hamann (8) and in some "data publications" cited later.

In the next section I will turn to consideration of publications that contain thermodynamic data that are useful for calculations involving aqueous solutions. It should also be mentioned that many of these "data publications" contain considerable information about uses of thermodynamic data for aqueous solutions.

SOURCES OF THERMODYNAMIC DATA

Because we are here concerned with aqueous solutions, it is appropriate to begin by citing two reviews (9,10) of the self-ionization of water that may be represented by

$$H_2O(\text{liq}) = H^+(\text{aq}) + OH^-(\text{aq}) \quad (17)$$

The best and most extensive sources of thermodynamic data for inorganic substances and for those organics containing only one or two carbon atoms per molecule have been published by the National Bureau of Standards (11,12). These excellent compilations are primarily concerned with thermodynamic data for 298.15 K (25.00°C), so it is necessary to look elsewhere for data that refer to other temperatures. It is also necessary to look elsewhere for the best available heat capacities for aqueous solutions, since many of these important quantities have been reported after the cutoff times for preparation of the NBS publications (11,12).

The best and most recent single-volume sources of thermodynamic data for inorganic solids (many of which can be involved in reactions with aqueous solutions) have come from the Geological Survey (13) and the Bureau of Mines (14).

Several compilations of thermodynamic properties of pure organic substances have been published (15-17). Many of these organic substances can be either reactants or products of reactions taking place in aqueous solutions.

The JANAF publications (18) contain much useful information, with a particular emphasis on high temperature thermodynamic data.

Latimer's classic "Oxidation Potentials" (19) is still a useful source of information about thermodynamics of aqueous solutions (especially for certain calculations and methods of estimating unknown quantities). Further information about oxidation-reduction (electrode) potentials is contained in other books (20,21).

Thermodynamic properties of various classes of substances (aqueous acids and bases, metal complexes) have been reviewed (22-27).

There have been many reviews (28-42) of selected thermodynamic properties of certain chemical elements and compounds. These reviews are mostly concerned with properties of aqueous species and also various gases and solids such as salts, oxides, and sulfides that can be involved in reactions or equilibria with aqueous solutions.

The reviews and sources of data cited so far have been mostly concerned with such thermodynamic properties as ΔG_f^o, ΔH_f^o, S^o, various equilibrium constants and related enthalpies of reaction, oxidation-reduction or electrode potentials, etc. Two textbooks (3,4) and one monograph (7) contain considerable information about activity coefficients of solutes in aqueous solution, which are needed for many calculations concerning concentrated solutions and for some calculations concerning dilute solutions. I also call attention to excellent reviews (43-52) from the National Bureau of Standards and original investigations (53-58) of specific systems by Pitzer and colleagues. Pitzer (59) has provided a review of recent advances in the theory of electrolyte solutions, with emphasis on activity coefficients and related theory that can be useful in connection with problems such as those under consideration in this paper. Finally, I call attention to chapters by Pitzer, Meissner, Staples, Friedman, and Pytkowicz in the important book (1) edited by Newman and to a paper by Pitzer and Murdzek (60) that is concerned with calculations involving activity coefficients, thermal properties, and solubilities of sodium sulfate at many temperatures.

McKay and Shiu (61) have provided a critical review of Henry's Law constants for a variety of substances that are of particular interest in connection with environmental problems. Several contributors to the book (1) edited by Newman have discussed the thermodynamics of gases dissolved in water and aqueous solutions. Related recent references (62-69) are cited here.

Partial molar volumes are useful in calculations on the effects of pressure on various properties as in Equations (15) and (16). We have from Millero (70,71) two reviews of such volumes for aqueous electrolytes. Five papers (72-75) published during 1982 provide more recent reviews and include some new results. Many other papers dealing with the volumetric properties of specific aqueous systems have been published in the last two decades by several of the authors of references (70-76) and also by the following: J.C. Ahluwalia, O.D. Bonner, G.A. Bottomley, S. Cabani, B.E. Conway, J.E. Desnoyers, J.-P.E. Grolier, H. Hoiland, C. Jolicoeur, F. Shahidi, R. Zana, and others.

Thermal properties (especially specific heats, partial molar heat capacities of solutes, and heats of dilution) are useful for a variety of thermodynamic calculations, as already illustrated for the specific calculation of the dependence of an equilibrium constant on temperature. Such properties have been reviewed through 1965 by Parker (77) and have been summarized in tabular form (only for 25°C) in recent publications (12) from the National Bureau of Standards. Because most of the best measurements of these properties (especially partial molar heat capacities of aqueous solutes) have been made since these reviews were prepared, it is necessary to search further for many of the data that are likely to be most useful. Smith-Magowan and Goldberg have prepared a useful bibliography (78) of sources of thermal data for aqueous solutions. In the few years since this bibliography was prepared many papers have appeared in which many such properties have been reported; examples are references 58, 60, 72, and 73 that have already been cited in another connection. Since it is impractical to cite here all of the other recent papers dealing with heats of dilution and especially with partial molar heat capacities of aqueous solutes, I merely list the following names of authors of many of these papers: J.C. Ahluwalia, G.C. Benson, O.D. Bonner, S. Cabani, J.W. Cobble, G. Conti, C.M. Criss, J.E. Desnoyers, S.J. Gill, J.-P.E. Grolier, L.G. Hepler, H.P. Hopkins, Jr., C. Jolicoeur, E. Matteoli, F.J. Millero, P. Picker, K.S. Pitzer, G. Somsen, P.R. Tremaine, I. Wadsö, R.H. Wood, E.M. Woolley, etc. A comprehensive review is needed and is being prepared by the author and colleagues.

In closing it is appropriate to call attention to two bibliographies (79,80) and to a recent review (81) of the thermodynamics of dilute solutions.

ACKNOWLEDGMENTS

I am grateful to the Natural Sciences and Engineering Research Council of Canada and to the Alberta Oil Sands Technology and Research Authorty for their support of my research on the thermodynamic properties of aqueous solutions. In addition, I thank Dr. R.N. Goldberg of the U.S. National Bureau of Standards for sending me a copy of his valuable listing of references to the literature dealing with thermodynamic properties.

LITERATURE CITED

1. Newman, S.A. (Ed.), H.E. Barner, M. Klein, and S.I. Sandler, (Assoc. Ed.), Thermodynamics of Aqueous Systems with Industrial Applications, ACS Symp. Series 133, Washington, D.C. (1980).

2. Lewis, G.N. and M. Randall, Thermodynamics and The Free Energy of Chemical Substances, McGraw-Hill, New York (1923).

3. Lewis, G.N. and M. Randall (revised by K.S. Pitzer and L. Brewer), Thermodynamics, 2nd ed., McGraw-Hill, New York (1961).

4. Klotz, I.M. and R.M. Rosenberg, Chemical Thermodynamics: Basic Theory and Methods, 3rd ed., W.A. Benjamin, Inc., Menlo Park, California (1972).

5. McGlashan, M.L., Chemical Thermodynamics, Academic Press, London (1979).

6. Denbigh, K., The Principles of Chemical Equilibrium, 4th ed., Cambridge University Press, Cambridge, England (1981).

7. Robinson, R.A. and R.H. Stokes, Electrolyte Solutions. The Measurement and Interpretation of Conductance, Chemical Potential and Diffusion in Solutions of Simple Electrolytes, 2nd ed., Butterworths, London (1965).

8. Hamann, S.D., Physico-Chemical Effects of Pressure, Butterworths, London (1957).

9. Olofsson, G. and L.G. Hepler, "Thermodynamics of Ionization of Water over Wide Ranges of Temperature and Pressure", J. Solution Chem. 4, 127 (1975).

10. Marshall, W.L. and E.U. Franck, "Ion Product of Water Substance, 0-1000°C, 1-10,000 Bars. New International Formulation and its Background", J. Phys. Chem. Ref. Data 10, 295 (1981).

11. Rossini, F.D., D.D. Wagman, W.H. Evans, L. Levine, and I. Jaffe, "Selected Values of Chemical Thermodynamic Properties", National Bureau of Standards Circular 500, U.S. Gov't Printing Office, Washington, D.C. (1952).

12. Wagman, D.D. and colleagues, "Selected Values of Chemical Thermodynamic Properties", NBS Technical Notes 270-1 through 270-8, U.S. Gov't Printing Office, Washington, D.C. (1965-1981). Similar tables have recently been published as: "NBS Tables of Chemical Thermodynamic Properties of Inorganic and C_1 and C_2 Organic Substances in SI Units, J. Phys. Chem. Ref. Data 11, Supplement 2 (1982).

13. Robie, R.A., B.S. Hemingway, and J.R. Fisher, "Thermodynamic Properties of Minerals and Related Substances at 298.15 K and 1 Bar (10^5 Pascals) Pressure and at Higher Temperatures", Geological Survey Bulletin 1452, U.S. Gov't Printing Office, Washington, D.C. (1978).

14. Pankratz, L.B., "Thermodynamic Properties of Elements and Oxides", Bureau of Mines Bulletin 672, U.S. Gov't Printing Office, Washington, D.C. (1982). This useful book also contains a section (by R.V. Mrazek) on Process Applications.

15. Stull, D.R., E.F. Westrum, Jr., and G.C. Sinke, The Chemical Thermodynamics of Organic Compounds, John Wiley and Sons, Inc., New York (1969).

16. Cox, J.D. and G. Pilcher, Thermochemistry of Organic and Organometallic Compounds, Academic Press, London (1970).

17. Janz, G.J., Thermodynamic Properties of Organic Compounds: Estimation Methods, Principles, and Practice, revised ed., Academic Press, New York (1967).

18. JANAF Thermochemical Tables, 2nd ed., NSRDS-NBS 37, U.S. Gov't Printing Office, Washington, D.C. (1971). Supplements have been published in J. Phys. Chem. Ref. Data 3, 311 (1974); 4, 1 (1975); 7, 793 (1978).

19. Latimer, W.M., The Oxidation States of the Elements and Their Potentials in

Aqueous Solution, 2nd ed., Prentice-Hall, Englewood Cliffs, New Jersey (1952).

20. Charlot, G., Selected Constants: Oxidation-Reduction Potentials of Inorganic Substances in Aqueous Solution, Butterworths, London (1971).

21. Clark, W.M., Oxidation-Reduction Potentials of Organic Systems, The Williams and Wilkins Co., Baltimore (1960).

22. Christensen, J.J., L.D. Hansen, and R.M. Izatt, Handbook of Proton Ionization Heats and Related Thermodynamic Quantities, John Wiley and Sons, New York (1976).

23. Larson, J.W. and L.G. Hepler, "Heats and Entropies of Ionization", in Solute-Solvent Interactions, Coetzee, J.F. and C.D. Ritchie, (Eds.), Marcel Dekker, New York (1969).

24. Hepler, L.G. and H.P. Hopkins, Jr., "Thermodynamics of Ionization of Inorganic Acids and Bases in Aqueous Solution", Rev. Inorg. Chem. 1, 303 (1979).

25. Baes, C.F., Jr., and R.E. Mesmer, The Hydrolysis of Cations, John Wiley and Sons, New York (1976).

26. Ashcroft, S.J. and C.T. Mortimer, Thermochemistry of Transition Metal Complexes, Academic Press, London (1970).

27. Christensen, J.J., D.J. Eatough, and R.M. Izatt, Handbook of Metal Ligand Heats and Related Thermodynamic Quantities, 2nd ed., Marcel Dekker, New York (1975).

28. Brewer, L., "Thermodynamic Values for Desulfurization Processes", in Flue Gas Desulfurization, ACS Symposium Series 188, Hudson, J.A. and G.T. Rochelle (Eds.) Washington, D.C. (1982).

29. Parker, V.B., B.R. Staples, T.L. Jobe, Jr., and D.B. Neumann, "A Report on Some Thermodynamic Data for Desulfurization Processes", NBSIR 81-2345, National Bureau of Standards, Washington, D.C. (1981).

30. Hepler, L.G. and G. Olofsson, "Mercury: Thermodynamic Properties, Chemical Equilibria, and Standard Potentials", Chem. Reviews 75, 585 (1975).

31. Gedansky, L.M. and L.G. Hepler, "Thermochemistry of Silver and Its Compounds", Engelhard Ind. Tech. Bull. IX, 117 (1969).

32. Gedansky, L.M. and L.G. Hepler, "Thermochemistry of Gold and Its Compounds", Engelhard Ind. Tech. Bull. X, 5 (1969).

33. Morss, L.R., "Thermochemical Properties of Yttrium, Lanthanum, and the Lanthanide Elements and Ions", Chem. Reviews 76, 827 (1976).

34. Travers, J.G., I. Dellien, and L.G. Hepler, "Scandium: Thermodynamic Properties, Chemical Equilibria, and Standard Potentials", Thermochim. Acta 15, 89 (1976).

35. Hepler, L.G. and P.P. Singh, "Lanthanum: Thermodynamic Properties, Chemical Equilibria, and Standard Potentials", Thermochim. Acta 16, 95 (1976).

36. Dellien, I., F.M. Hall, and L.G. Hepler, "Chromium, Molybdenum, and Tungsten: Thermodynamic Properties, Chemical Equilibria, and Standard Potentials", Chem. Reviews 76, 283 (1976).

37. Zordan, T.A. and L.G. Hepler, "Thermochemistry and Oxidation Potentials of Manganese and Its Compounds", Chem. Reviews 68, 737 (1968).

38. Wagman, D.D., R.H. Schumm, and V.B. Parker, "A Computer-Assisted Evaluation of the Thermochemical Data of the Compounds of Thorium", NBSIR 77-1300, National Bureau of Standards, Washington, D.C. (1977).

39. Parker, V.B., "The Thermochemical Properties of the Uranium-Halogen Containing Compounds", NBSIR 80-2029, National Bureau of Standards, Washington, D.C. (1980).

40. Goldberg, R.N. and L.G. Hepler, "Thermochemistry and Oxidation Potentials of the Platinum Group Metals and Their Compounds", Chem. Reviews 68, 229 (1968).

41. Hill, J.O., I.G. Worsley, and L.G. Hepler, "Thermochemistry and Oxidation Potentials of Vanadium, Niobium, and Tantalum", Chem. Reviews 71, 127 (1971).

42. Gedansky, L.M., E.M. Woolley, and L.G. Hepler, "Thermochemistry of Compounds and Aqueous Ions of Copper", J. Chem. Thermodynamics 2, 561 (1970).

43. Goldberg, R.N., B.R. Staples, and R.L. Nuttall, "Evaluated Activity and Osmotic Coefficients for Aqueous Solutions: Iron Chloride and the Bi-Univalent Compounds of Nickel and Cobalt", J. Phys.

Chem. Ref. Data 8, 923 (1979).

44. Goldberg, R.N., "Evaluated Activity and Osmotic Coefficients for Aqueous Solutions: Bi-Univalent Compounds of Lead, Copper, Manganese, and Uranium", J. Phys. Chem. Ref. Data 8, 1005 (1979).

45. Goldberg, R.N., "Evaluated Activity and Osmotic Coefficients for Aqueous Solutions: Bi-Univalent Compounds of Zinc, Cadmium, and Ethylene Bis(Trimethylammonium) Chloride and Iodide", J. Phys. Chem. Ref. Data 10, 1 (1981).

46. Goldberg, R.N., "Evaluated Activity and Osmotic Coefficients for Aqueous Solutions: Thirty-Six Uni-Bivalent Electrolytes", J. Phys. Chem. Ref. Data 10, 761 (1981).

47. Staples, B.R., "Activity and Osmotic Coefficients of Aqueous Metal Nitrites", J. Phys. Chem. Ref. Data 10, 765 (1981).

48. Staples, B.R., "Activity and Osmotic Coefficients of Aqueous Sulfuric Acid", J. Phys. Chem. Ref. Data 10, 779 (1981).

49. Staples, B.R. and R.L. Nuttall, "The Activity and Osmotic Coefficients of Aqueous Calcium Chloride at 298.15 K", J. Phys. Chem. Ref. Data 6, 385 (1977).

50. Goldberg, R.N. and R.L. Nutall, "Evaluated Activity and Osmotic Coefficients for Aqueous Solutions: The Alkaline Earth Metal Halides", J. Phys. Chem. Ref. Data 7, 263 (1978).

51. Hamer, W.J. and Y.C. Wu, "Osmotic Coefficients and Mean Activity Coefficients of Uni-Univalent Electrolytes in Water at 25°C", J. Phys. Chem. Ref. Data 1, 1047 (1972).

52. Wu, Y.C. and W.J. Hamer, "Electrochemical Data. Part XIV. Osmotic Coefficients and Mean Activity Coefficients of a Series of Uni-Bivalent and Bi-Univalent Electrolytes in Aqueous Solution at 25°C. Part XVI. Osmotic Coefficients and Mean Activity Coefficients of a Number of Uni-Trivalent and Tri-Univalent Electrolytes in Aqueous Solution at 25°C", NBSIR No. 10052 and 10088, National Bureau of Standards, Washington, D.C. (1969).

53. Pitzer, K.S., "Thermodynamic Properties of Aqueous Solutions of Bivalent Sulphates", J. Chem. Soc. Faraday Trans. II 68, 101 (1972).

54. Pitzer, K.S. and J.J. Kim, "Thermodynamics of Electrolytes. IV. Activity and Osmotic Coefficients for Mixed Electrolytes", J. Am. Chem. Soc. 96, 5701 (1974).

54a. Pitzer, K.S., R.N. Roy, and L.F. Silvester, "Thermodynamics of Electrolytes. VII. Sulfuric Acid", J. Am. Chem. Soc. 99, 4930 (1977).

55. Silvester, L.F. and K.S. Pitzer, "Thermodynamics of Electrolytes. 8. High-Temperature Properties, Including Enthalpy and Heat Capacity, with Application to Sodium Chloride", J. Phys. Chem. 81, 1822 (1977).

56. Silvester, L.F. and K.S. Pitzer, "Thermodynamics of Electrolytes. X. Enthalpy and the Effect of Temperature on the Activity Coefficients", J. Solution Chem. 7, 327 (1978).

57. Pitzer, K.S. and J.C. Peiper, "Activity Coefficients of Aqueous $NaHCO_3$", J. Phys. Chem. 84, 2396 (1980).

58. Peiper, J.C. and K.S. Pitzer, "Thermodynamics of Aqueous Carbonate Solutions Including Mixtures of Sodium Carbonate, Bicarbonate, and Chloride", J. Chem. Thermodynamics 14, 613 (1982).

59. Pitzer, K.S., "Electrolyte Theory - Improvements Since Debye and Huckel", Accounts Chem. Res. 10, 371 (1977).

60. Pitzer, K.S. and J.S. Murdzek, "Thermodynamics of Aqueous Sodium Sulfate", J. Solution Chem. 11, 409 (1982).

61. MacKay, D. and W.Y. Shiu, "A Critical Review of Henry's Law Constants for Chemicals of Environmental Interest", J. Phys. Chem. Ref. Data 10, 1175 (1981).

62. Schulze, G. and J.M. Prausnitz, "Solubilities of Gases in Water at High Temperatures", Ind. and Eng. Chem. Fund. 20, 175 (1981).

63. Cramer, S.D., "The Solubility of Methane, Carbon Dioxide, and Oxygen in Brines from 0° to 300°C", Bur. Mines Rep. Inv. 8706 (1982).

64. Tochigi, K. and K. Kojima, "Prediction of Non-Polar Gas Solubilities in Water, Alcohols and Aqueous Solutions by the Modified Asog Method", Fluid Phase Equil. 8, 221 (1982).

65. Benson, B.B. and D. Krause, Jr., "Empirical Laws for Dilute Aqueous Solutions of Nonpolar Gases", J. Chem. Phys. 64, 689 (1976).

66. Lee, J.I., F.D. Otto, and A.E. Mather, "Solubility of Carbon Dioxide in Aqueous Diethanolamine Solutions at High Pressures", J. Chem. Eng. Data 17, 465 (1972).

67. Potter, R.W., II and M.A. Clynne, "The Solubility of the Noble Gases He, Ne, Ar, Kr, and Xe in Water up to the Critical Point", J. Solution Chem. 7, 837 (1978).

68. O'Sullivan, T.D. and N.O. Smith, "The Solubility and Partial Molar Volume of Nitrogen and Methane in Water and in Aqueous Sodium Chloride from 50 to 125° and 100 to 600 Atm", J. Phys. Chem. 74, 1460 (1970).

69. Gardiner, G.E. and N.O. Smith, "Solubility and Partial Molar Properties of Helium in Water and Aqueous Sodium Chloride from 25 to 100°C and 100 to 600 Atmospheres", J. Phys. Chem. 76, 1195 (1972).

70. Millero, F.J., "The Partial Molal Volumes of Electrolytes in Aqueous Solutions", in Water and Aqueous Solutions, Horne, R.A. (Ed.), John Wiley and Sons, Interscience, New York (1972).

71. Millero, F.J., "The Molal Volumes of Electrolytes", Chem. Reviews 71, 147 (1971).

72. French, R.N. and C.M. Criss, "Effect of Charge on the Standard Partial Molar Volumes and Heat Capacities of Organic Electrolytes in Methanol and Water", J. Solution Chem. 11, 625 (1982).

73. Larson, J.W., K.G. Zeeb, and L.G. Hepler, "Heat Capacities and Volumes of Dissociation of Phosphoric Acid, (1st, 2nd, and 3rd) Bicarbonate Ion, and Bisulfate Ion in Aqueous Solution", Can. J. Chem. 60, 2141 (1982).

74. Leyendekkers, J.V., "Ionic Contributions to Partial Molal Volumes in Aqueous Solutions", J. Chem. Soc. Faraday Trans. I 78, 357 (1982).

75. Millero, F.J., "The Effect of Pressure on the Solubility of Minerals in Water and Seawater", Geochim. Cosmochim. Acta 46, 11 (1982).

76. Moore, J.C., R. Battino, T.R. Rettich, Y.P. Handa, and E. Wilhelm, "Partial Molar Volumes of 'Gases' at Infinite Dilution in Water at 298.15 K", J. Chem. Eng. Data 27, 22 (1982).

77. Parker, V.B., "Thermal Properties of Aqueous Uni-Univalent Electrolytes", NSRDS NBS 2, U.S. Gov't Printing Office, Washington, D.C. (1965).

78. Smith-Magowan, D. and R.N. Goldberg, "A Bibliography of Sources of Experimental Data Leading to Thermal Properties of Binary Aqueous Electrolyte Solutions", NBS Special Publication 537, U.S. Gov't Printing Office, Washington, D.C. (1979).

79. Goldberg, R.N., B.R. Staples, R.L. Nuttall, and R. Arbuckle, "A Bibliography of Sources of Experimental Data Leading to Activity or Osmotic Coefficients for Polyvalent Electrolytes in Aqueous Solution", NBS Special Publication 485, U.S. Gov't Printing Office, Washington, D.C. (1977).

80. Hawkins, T., "A Bibliography on the Physical and Chemical Properties of Water: 1969-1974", J. Solution Chem. 4, 625 (1975). Properties of dilute solutions as well as of pure water are covered.

81. Hepler, L.G., "Thermodynamics of Dilute Solutions", Pure & Applied Chem. 55, 493 (1983).

THE GAS PROCESSORS ASSOCIATION THERMODYNAMIC DATA FOR THE GAS PROCESSING INDUSTRY THROUGH COOPERATIVE RESEARCH

David F. Bergman ∎ Amoco Production Company, Tulsa, Oklahoma
Carl Sutton ∎ Gas Processors Association, Tulsa, Oklahoma

The Gas Processors Association's cooperative research program is directed at obtaining thermodynamic data for improving process designs. Research topics have expanded in scope in recent years to include substitute gases and production gases from EOR projects in addition to natural gas systems. The mechanics of this cooperative research effort is discussed along with the description of past and present research projects.

The Gas Processors Association was founded in 1921 as a trade association of natural gasoline manufacturers. Its initial objective was to develop and adopt standard specifications for natural gasoline. Today the GPA is a nonincorporated, nonprofit association made up of over 250 corporate members who account for nearly 90% of all natural gas liquids produced in the United States. Included in this total are 47 member companies based outside the United States. GPA moved into the area of thermodynamic data research in 1955 in order to develop an improved K-value correlation for light hydrocarbons. Most people are familiar with the NGAA K-Value Charts that were prepared from this study. However, the GPA is much more than a research organization. It develops standards and specifications for the gas processing industry and is presently involved with the International Standards Organization to develop international standards. It holds technical seminars and an annual convention. Finally, it provides technical input to Government agencies to help in the writing of regulations. The following discussion will center on the cooperative research effort of the GPA.

Development of Research Topics

Most of the research projects are administered by the Technical Data Committee or, more precisely, Section F of that Committee. Section F is further divided into the K-Value and Enthalpy steering committees to facilitate more specialized attention to these research areas. Primary responsibility for initiation and supervision of research topics is vested in these two steering committees. This responsibility has grown in recent years as the budget has increased significantly. Presently over $4 million has been spent on research with yearly budgets now exceeding $500,000. Figure 1 shows the yearly allocations since 1963. The steering committees are comprised of experienced technical specialists who monitor the status, availability, and reliability of thermodynamic and physical data needed for design and operation of gas processing facilities. Figure 2 lists the present membership of these two steering committees. These steering committees meet twice a year to discuss research projects and prepare the research proposal for the coming years. The following list of criteria are applied in the development of all research projects.

1. Research must be directed toward real industry problems where solutions are considered possible.

2. Research must concentrate on areas of major importance to the gas processing industry.

3. The dollar value to the industry must be evaluated in advance.

4. Research programs must avoid proprietary or patent problems.

Additionally, the collection of experimental data has been emphasized over model development. The reasoning behind this is that there are many individuals who will develop models once the data are available. With this in mind, research projects have been directed towards obtaining a limited set of experimental data over a wide range of temperature, pressure and composition to serve as a basis for model development and improvement. In a few cases where there was an impending need for a model that was not forthcoming from the modelers, the GPA has appropriated funds for that purpose. Presently the GPA is marketing 8 computer programs, all but one developed with GPA funds. Figure 3 lists the programs with a brief description.

Mechanics of GPA Research

Once the initial step of outlining a project has been completed by one of the steering committees, the approval and funding process begins. A project description is prepared which defines the scope and justification for the project, including the estimated life span and overall cost. Then a joint proposal from both steering committees is sent to the parent committee, Section F. This proposal will contain 10-20 projects and a budget for the duration of the project. With Section F approval, the proposal is sent on to the Technical Committee. Since the Technical Committee represents all phases of GPA technical interest, it provides valuable screening for projects which lack general interest. Once approved by the Technical Committee, the yearly proposal is presented to the Board of Directors who must approve each project by a 75% majority. Approval here authorizes the project to be placed on a research ballot that will be submitted to the membership for funding approval. It is generally implied that multiyear projects, once initiated, will continue to receive funding until the project is completed. Year-to-year approval is still required, however. Occasionally the life span of a project will be changed to reflect industry needs and availability of funds.

Funding Approval

Final authorization is granted by GPA member companies who vote individually on each separate project of a yearly proposal. Usually there will be 10 to 20 projects proposed in any given year, with one or two failing to be approved. Figure 4 details the 1983 Proposal that was sent to the member companies for approval. The votes of member companies are weighted by their natural gas liquids (NGL) production. A project must receive a favorable vote from companies owning 75% of the total NGL production of GPA membership. For each approved project, the supporting companies are assessed in the proportion that their production bears to the total production from all supporting companies. Thus, if Company A supports a project that receives 80% approval and Company A's NGL production is 10% of the total production of supporting companies, Company A will be assessed 10% of the project cost for that year. Company A need not continue to support the project for more than one year if it so desires. For those projects failing to get the 75% favorable vote, a supplemental ballot of those companies originally supporting the project will sometimes be initiated. Here the supporting companies are asked to increase their support if they want this project to be funded. Figure 5 outlines the approval process.

With all of the approvals completed, a member of the steering committee where a project originated must prepare a request for proposal (RFP). The RFP outlines the work anticipated and funds available. Usually a temperature-pressure grid and composition will also be specified. The RFP's are then sent out to appropriate investigators who have the specialized equipment needed for the work. At the next Steering Committee meeting, an investigator is chosen from those responding to the RFP. Final approval from the Technical Committee is then required for each new investigator on a project. With this approval, the funds are released and work begins.

One member of a steering committee is selected for liaison with each investigator. It is his responsibility to maintain contact with the investigator and distribute progress reports. The liaison work sometimes requires the recontacting of committee members to discuss changes in a project.

Presentation of Results

Beginning in 1971, formal research reports were prepared by the investigator and distributed to all members and participating companies. These formal reports are prepared at appropriate points in multiyear projects or at the end of a shorter term project. Presentation of raw experimental data is a prerequisite. Reporting of smooth data is optional. Any interpretation of results and experimental detail that will improve future understanding of the data are included as appropriate. Presently there have been over 70 research reports completed and distributed. These reports are distributed to all member companies whether or not they participated and funded a particular project. Occasionally these projects are jointly funded. Currently Project 826 on CO_2 is receiving additional industry support. Research Report RR62, "Water Hydrocarbon Liquid-Liquid Vapor Equilibrium Measurement to 530°F", was a joint research report with the American Petroleum Institute. GPA officers and committee members keep in touch with other cooperative research organizations to promote joint funding where possible. This makes everyone's dollar go further and insures widespread distribution of the data.

1983 Research Program

The following is a list of the 1983 research projects and a very brief description of the area of research. This list is taken directly from the 1983 research report prepared by the GPA. Fifteen projects were approved for this year.

I. ENTHALPY RESEARCH

A. Natural Gas Systems

Project 792 - Experimental Enthalpies on Low BTU Gas
Objective: To extend the enthalpy data base for very sour gas systems to higher pressures. Data on pure toluene and methylcyclohexane will be taken prior to mixture data.

Project 811 - Experimental Enthalpies of Hexane + Fractions
Objective: To obtain experimental enthalpies on aromatic, naphthenic, and paraffin compounds normally found in the C_6+ fraction of natural gas streams. Data on synthetic mixtures containing the three pure compound types, as well as data on an actual hexane + fraction from a natural gas stream, will be measured.

Project 821 - Enthalpies of Acid Gas Absorption in Solvents
Objective: To obtain the enthalpies of absorption of acid gases in a variety of liquid solvents (e.g., TEA, MEA, ...).

Project 831 - Ideal Gas Enthalpies of Petroleum Fractions
Objective: To obtain ideal gas data by employing modern calculational tools. This data will be used to extend the data base to higher temperatures for petroleum fractions.

B. Supplemental Gas Systems

No approved projects for 1983.

C. General

Project 822/806 - Data Handling and Evaluation
Objective: To establish a data bank for evaluation of a broad range of enthalpy and phase equilibrium models presently available. Continuing funds will be used to update the data base and evaluate new models as they become available.

II. PHASE EQUILIBRIUM RESEARCH

A. Natural Gas Systems

Project 755 - VLE Data for Modeling Non-paraffinic Hydrocarbons
Objective: To obtain binary experimental phase equilibrium data on nonhydrocarbon components, N_2, CO_2, and H_2S, with components of natural gas.

Project 795 - VLL/VLS Equilibria for Cryogenic Processing
Objective: To obtain experimental data in the vapor-liquid-liquid region and vapor-liquid-solid region and then to use these data for model development for prediction of solid and liquid-liquid formation in cryogenic gas processing plants.

Project 815 - Data for Evaluation of Phase Equilibria Models
Objective: To obtain vapor-liquid equilibrium ratios, phase boundaries and liquid phase fractions in the dewpoint or retrograde region.

Project 825 - VLL Equilibria for Gas-Water Systems with Dehydration Fluids
Objective: To obtain equilibrium measurements on natural gas systems containing N_2, CO_2, and H_2S with methanol in the hydrate region. Future work will include other alcohols and/or glycols.

B. Enhanced Recovery Systems

Project 826 - Processing Gas Associated with EOR Projects
Objective: To obtain vapor liquid equilibria data on systems containing CO_2 in concentrations similar to those expected during distillation to separate CO_2 from produced gases.

C. Hydrocarbon Water Systems

Project 758 - Data and Modeling of Water with Gaseous Components
Objective: To obtain phase equilibrium data for water with natural gas streams to be used in the development of the necessary interaction parameters to predict the behavior of the water.

Project 775 - Water Content of Gas and NGL in the Presence of Hydrate
Objective: To obtain data on the hydrate boundary of gas streams below the freezing point of water and inside the hydrate phase region.

D. Supplemental Gas Systems

No projects approved for 1983

E. Sour Water

Project 835 - Data for Removal of Sour Components from Water
Objective: To extend binary data obtained in Project 758 and Project 805 by focusing on the effect of ammonium chloride and phenol on sour water equilibria.

III. MEASUREMENT RESEARCH

Project 829 - Update of AGA Basin Orifice Flow Coefficients (Co-sponsored with the American Petroleum Institute)
Objective: To obtain new data to correct the errors contained in AGA Report No. 3 published in 1934.

SUMMARY

Cooperative research has been an excellent way for supporting experimental research in areas of general interest. Not only does is lower the cost, but it allows smaller companies to obtain results from a major research project where it could not have justified a large expenditure. Cooperative research also avoids duplication of effort since many companies will be faced with the same data needs. Funding this effort helps support research at many universities which are undergoing financial crises at the present time and need an additional source of funds. There are some negatives, however, to cooperative research. Since much of the experimental work is carried out at universities, it takes longer to be completed. This is generally due to a limited school year and also to more frequent changing of personnel than would be experienced at a research laboratory. Additionally, a company might find itself supporting a project that either doesn't do enough or does more than necessary to satisfy its particular needs. However, a critical evaluation of the experimental data that has been made available to the industry, shows that the cooperative research effort at the GPA has been very beneficial to the gas processing industry.

NOTE: A list of research reports published through the GPA can be obtained by writing to the GPA Office:

1812 First Place
15 E. Fifth Street
Tulsa, OK 74103

Date	GPA Funds	Non-Member Funds	Total
1963-1970	$747,000	$80,000	$ 827,000
1971	120,000	20,000	140,000
1972	96,000	17,000	113,000
1973	103,000	13,000	116,000
1974	151,000	20,000	171,000
1975	168,000	30,000	198,000
1976	172,000	30,000	202,000
1977	200,000	30,000	230,000
1978	248,000	26,000	274,000
1979	222,000	42,000	264,000
1980	213,500	29,000	242,500
1981	335,000	30,000	365,000
1982	470,000	30,000	500,000
TOTAL 1963-1982			$3,642,500
Proposed for 1983	$610,000	$50,000	$ 660,000

Figure 1. Approved research project allocations.

M. A. Albright, Section F Chairman, Phillips Petroleum Co., Bartlesville

Enthalpy Steering.
Chairman: L. D. Wiener, Mobil Research & Development Corp., Dallas
  F. G. Clark, C F Braun & Co., Alhambra, Calif.
  Bill A. Alderman, Cities Service Co., Tulsa
  Joe Provine, Conoco Inc., Ponca City
  Monica Baltatu, Fluor E & C, Irvine, Calif.
  Stephen A. Newman, Foster Wheeler Energy, Livingston, N.J.
  Pat Campbell, McDermott Marine Engineering, Houston
  R. R. Wood, Shell Development Co., Houston

Phase Equilibrium Steering.
Chairman, Karl Kilgren, Chevron Research Co., Richmond, Calif.
  David Zudkevitch, Allied Chemical Corp., Morristown, N.J.
  David F. Bergman, Amoco Production Co., Tulsa
  Allen M. Rowe, Arco Oil & Gas Co., Dallas
  Douglas G. Elliot, DM International Inc., Houston
  David Rule, Exxon Production Research, Houston
  Hans F. Kistenmacher, Linde AG, Munich, Germany
  Robert L. McKee, Matthew Hall Engineering, Houston
  B. I. Lee, Mobil R & D Corp., Princeton, N.J.
  Pervaiz Nasir, Shell Oil Co., Houston
  John E. Johnson, Stearns-Roger Corp., Denver

Figure 2. Technical section F roster.

GPA * SIM

This program is an outgrowth of an original GPA programming of the Chao-Seader correlation and the GPA K&H flash program. However, numerous modifications by Professor John Erbar, Oklahoma State University, have extended its application.

Capable of handling either two- or three-phase calculations, it utilizes the Soave Redlich Kwong (SRK) equation. It is applicable to common light hydrocarbon mixtures, including common non-hydrocarbon components hydrogen, nitrogen, oxygen, carbon monoxide, carbon dioxide, hydrogen sulfide, and water.

This program is designed to operate in an interactive mode or batch operation for in-house systems. It can also be accessed through time-share systems of General Electric in the U.S., Canada, U.K., Europe, and Australia and University Computing Co. in the U.S.

Equi-Phase

This is a comprehensive set of phase behavior computer programs for the hydrocarbon industry developed by Drs. D. Y. Peng, H. J. Ng and Don Robinson, formerly with the University of Alberta. It consists of two main parts: the phase behavior and fluid properties package and the hydrate package. The programs are applicable wherever hydrocarbon mixtures are being produced, transported or processed. All necessary parameters for treating hydrocarbons from methane through heavy oils and H_2O, N_2, H_2, H_2S and CO_2 are built into the program. Recently developed features include hydrate inhibitor calculations using alcohols, glycols and salt.

GPA Cryogenic Solubility

Professors James A. Kohn, University of Notre Dame, and Kraemer Luks, University of Tulsa, have developed data to predict the solubility of CO_2 and hydrocarbons in cryogenic LNG and NGL. The program offers a solution to the multicomponent S-L-V equilibrium problem pertinent to the formation of solids in LNG and NGL systems. It is designed to take a feed stream composition and either (a) analyze for the possible formation of solids in the liquefied feedstream at a given temperature; or (b) determine the maximum temperature at which solids could exist in the liquefied feed stream. Standard components are C_1 through C_{10}, benzene, cyclohexane, carbon dioxide and nitrogen.

GPA TRAPP

The TRAPP program calculates the viscosity, thermal conductivity, and density of natural gas and other hydrocarbon mixtures. Range of applicability is 0-15,000 psia and $-300°$ to $450°$ F. Accuracy of prediction falls within $\pm 15\%$ of theoretical norm. TRAPP was developed by Drs. J. F. Ely and H. J. M. Hanley with the Thermophysical Properties Division of National Bureau of Standards, Boulder, Colo., and is available only to member companies through GPA. Non-GPA members should contact: Office of Standard Reference Data, National Bureau of Standards, Washington, D.C., 20234, (301) 921-2467.

GPA-OSU Sour Gas Program

Developed by Prof. R. N. Maddox at Oklahoma State University, Stillwater, this program makes the calculations necessary to predict the equilibrium between acid gas constituents and ethanolamine treating solutions. The program predicts vapor partial pressures or solution loadings for either pure H_2S, pure CO_2, or mixtures of H_2S and CO_2 in natural gas systems. It can predict equilibrium conditions for monoethanolamine, diethanolamine, di-isopropanolamine, and diglycolamine. As equilibrium data for additional ethanolamine-acid gas systems become available, they will be incorporated into the program.

GPA/NASA Chemical Equilibria

This program was developed from an earlier, larger NASA program to establish chemical equilibrium data directed toward synthetic gas calculations to solve typical combustion problems.

GPA Liquid Properties 80

This program will estimate bulk thermodynamic properties of light hydrocarbon liquid mixtures and may be incorporated into the user's own process program. It will calculate liquid density, enthalpy, entropy, and fugacity for specified temperature, pressure, and composition. Standard components are C_1 through C_{10}, carbon dioxide, hydrogen sulfide and nitrogen.

K-Data Program

Curve fitted coefficients for K-value charts appearing in the GPSA Engineering Data Book.

Figure 3. GPA computer programs.

Project No.[1]	Description of Work	Source of Funds			Totals
		GPA	Internal Sources[2]	Non-Members[3]	
	I. Enthalpy Research				
	A. Natural Gas Systems				
792	Experimental Enthalpies of Low Btu Gas	$ 20,900	$ 6,800	$ 2,300	$ 30,000
811	Experimental Enthalpies of Hexanes-plus Fractions	31,400	10,200	3,400	45,000
821	Enthalpies of Acid Gas Absorption in Solvents	31,400	10,200	3,400	45,000
831	Ideal Gas Enthalpies of Petroleum Fractions	20,900	6,800	2,300	30,000
	B. Supplemental Gas Systems				
773	Experimental Enthalpies for Binary Gas Systems	41,900	13,600	4,500	60,000
	C. General				
822/806	Data Handling & Evaluation	6,900	2,300	800	10,000
	Contingency Fund	6,900	2,300	800	10,000
				Enthalpy Research Total	**$230,000**
	II. Phase Equilibrium Research				
	A. Natural Gas Systems				
755	VLE Data for Modeling Nonparaffinic Hydrocarbons	$ 27,900	$ 9,100	$ 3,000	$ 40,000
795	V-L-L Equilibria for Cryogenic Processing	31,400	10,200	3,400	45,000
815	Data for Evaluation of Phase Equilibrium Models	20,900	6,800	2,300	30,000
825	V-L-L Equilibria for Gas-Water Systems with Dehydration Fluids	24,500	8,000	2,500	35,000
	B. Enhanced Recovery Systems				
826	Processing Gas Associated with EOR Projects	34,800	11,400	3,800	50,000
	C. Hydrocarbon-Water Systems				
758	Data & Modeling of Water with Gaseous Components	20,900	6,800	2,300	30,000
775	Water Content of Gas & NGL in Presence of Hydrate	34,800	11,400	3,800	50,000
	D. Supplemental Gas Systems				
757	VLE Data & Modeling of Supplemental Gas Components	20,900	6,800	2,300	30,000
	E. Sour Water				
835	Data for Removal of Sour Components from Water	27,900	9,100	3,000	40,000
	F. General				
806/822	Data Handling & Evaluation	6,900	2,300	800	10,000
	Contingency Fund	6,900	2,300	800	10,000
				Phase Equilibrium Research Total	**$370,000**
	III. Measurement Research				
829	Update of AGA Basic Orifice Flow Coefficients	$ 41,900	$ 13,600	$ 4,500	$ 60,000
	Total 1983 Research Budget	**$460,000**	**$150,000**	**$50,000**	**$660,000**

[1] First two digits of project numbers indicate initial year of project. Numbers 1–4 for the last digit indicate Enthalpy projects; 5–8 indicate Phase Equilibrium projects; 9 indicates general interest project.
[2] Internal Sources are computer program sales and interest on temporarily idle funds.
[3] Non-Member Participants include engineering contractors, supply firms, research organizations, etc.

Figure 4. Proposed 1983 research budget.

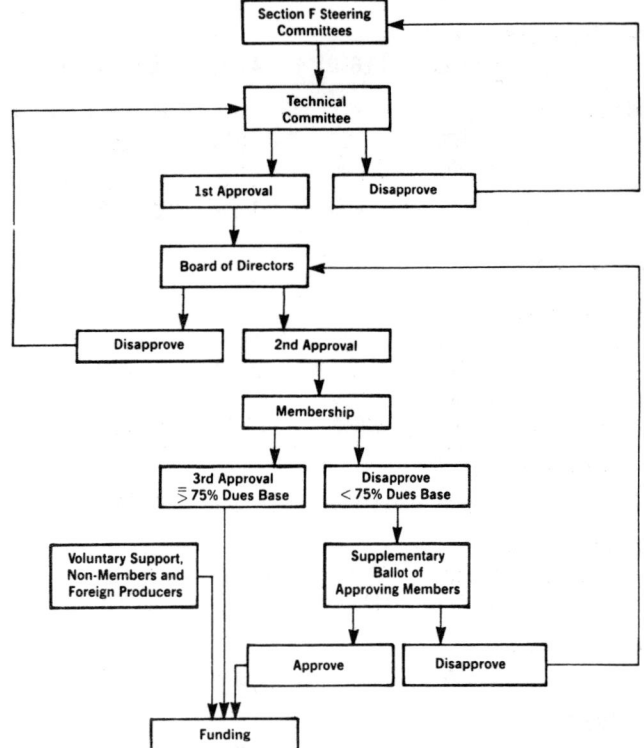

Figure 5. Project approval and funding flo-chart.

THERMOPHYSICAL PROPERTY DATA GENERATED BY THE NBS CENTER FOR CHEMICAL ENGINEERING

N.A. Olien ■ Chemical Engineering Science Division, Center for Chemical Engineering, National Bureau of Standards, Boulder, CO

H.J. Raveché ■ Thermophysics Division, Center for Chemical Engineering, National Bureau of Standards, Washington, DC

This paper describes the current and recent past research of the National Bureau of Standards (NBS) Center for Chemical Engineering (CCE) in the area of thermophysical properties. Included is a description of the approach used which integrates experimental, theoretical, and data evaluation efforts. There is also a summary of the impact of data and its accuracy on the chemical process industry. The major portion of the paper is a detailed description of the property research, especially the publications, of the Center over the past ten years. The results are presented in four tables, listing references of importance in the following areas: Pure Fluid Data, Fluid Mixture Data, Handbooks, Bibliographies, Computer Programs, and Solid Property Data. The paper concludes with a brief description of the future direction to be taken by the NBS-Center for Chemical Engineering in thermophysical properties.

The enabling congressional legislation which established the National Bureau of Standards at the beginning of this century serves to define the Laboratory's roles as a national resource in particular areas of science and technology. The mission of the NBS includes the development of state-of-the-art measurement methods, internationally accepted standards by which measurements can be appraised and critically evaluated reference data to be used by scientists and engineers throughout the nation's academic and industrial research laboratories. As part of this mission, the provision of high quality thermophysical property data on technically important fluids and solids has become an important function of NBS. In order to advance NBS work in the area of chemical engineering, several NBS Divisions were brought together to form the Center for Chemical Engineering in 1980. Although NBS has conducted much research in support of chemical engineering, the formation of the Center established for the first time a focus for this work. In response to widespread demand, property data are a significant part of the research output of this Center. Figure 1 illustrates the constituent parts of the Center. The Chemical Engineering Science Division and the Thermophysics Division have staff who are internationally recognized for providing state-of-the-art measurements and data for thermophysical properties, and both Divisions have been closely associated with the NBS-Office of

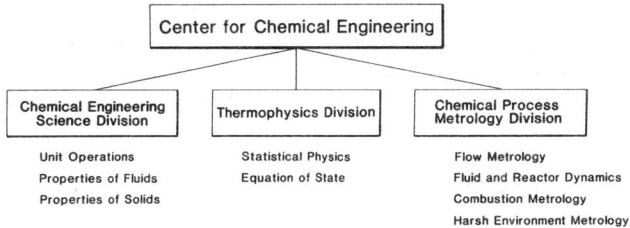

Figure 1. Organization of the NBS-Center for Chemical Engineering.

Standard Reference Data since the latter was formed in 1963.

The purpose of this paper is to summarize research activities in the area of thermophysical property data in the NBS-CCE and to discuss the importance of these data.

IMPORTANCE OF DATA

The importance of data in the chemical process industry has long been recognized in the design and operation of plants and facilities, especially in the area of custody transfer. The significance and economic impact of the accuracy of data, particularly the assignment of error bars to property values, has not been widely recognized or studied. This situation is changing rapidly, however, and quantitative studies of the effects of data inaccuracy have appeared recently. A pioneering paper was written by Ovid Baker in 1960 (1), David Zudkevitch

(2,3,4) has been an ardent advocate of the importance of accuracy in the thermophysical property base. The importance of these data bases has also been highlighted by Mel Albright (5), K. C. Chao (6), Max Klein (7) Joseph Kestin (8), Neil Olien (9), Doug Elliot (10), and Miller and Geist (11). In addition, there have been several studies to determine critical data needs for the future in specific technologies; see for example (12,13). The latter report points out that many future technologies will require property values for which there is no data base, let alone an accurate data base. For example, future technologies involved in alternative sources of chemical feedstocks, synthetic fuels, biotechnology, etc., will involve very complex fluid mixtures and heterogeneous solids for which thermophysical properties are both scarce and poorly understood.

THE NBS APPROACH TO THERMOPHYSICAL PROPERTIES

As the complexity and variability of mixtures increases, the possibility and desirability of making enough measurements to provide the needed data base declines rapidly. An approach is needed which consists of a judicious integration of experimental measurements on carefully selected systems, the critical evaluation of data, and theoretical studies all of which are geared to providing validated predictive techniques. The goal in developing accurate predictive techniques is to provide methods which are capable of handling a large number of chemical constituents and broad ranges of pressure, temperature, and composition. Such research should go beyond the correlation of data to include the prediction of data. This is the approach used by both the Thermophysics Division and the Chemical Engineering Science Division. To make these methods widely available to users in a convenient form, the results of our property research, particularly predictive techniques, take the form of validated computer programs. This is not done at the exclusion of tabular data, because many users have need of tables. In the case of property data for mixtures and complex solids, however, tables are just not practical. It is important that computer packages include extensive documentation and that confidence limits are clearly identified. The computer packages should also include provisions for notifying the user when calculations are requested which are outside the range of validity of the computer programs.

It is clear that predictive techniques in the form of validated computer programs are rapidly becoming the preferred form of thermophysical property data. This trend is being accelerated with the very recent increase in word size and therefore, numerical precision, of small computers.

DATA FROM THE NBS-CENTER FOR CHEMICAL ENGINEERING

The summary of published work on thermophysical properties data will be discussed in three sections: pure fluids, mixtures, and solids.

Pure Fluid Data. Accurate data for pure fluids are useful in several respects: for selected processes and for custody transfer, and as the major input to mixture predictive techniques. In an integrated approach such as described above, data for some fluids are vital to the development of mixtures models but of little importance as pure fluid data, while data for other pure fluids such as ethylene are important in their own right.

Table 1 lists the principal publications from the NBS-Center for Chemical Engineering in recent years on pure fluid properties. In many cases, these publications were preceded by others which present the basic documentation for the experimental measurements and data evaluation, but the listed publications are the most useful for engineering purposes.

Fluid Mixture Data. We previously stated that for the most part it is not particularly useful to present mixture data, especially multicomponent mixture data, in tabular form. Data for binary mixtures and other well-characterized systems are useful, in fact vital, to the development of predictive techniques. Phase equilibria data are useful per se and a substantial amount of accurate liquid-vapor equilibrium (LVE) data exist. This is not the case for single phase data where there is a dearth of accurate wide-range PVTx data for binary mixtures.

Table 2 lists the principal sources of our mixture data. The listed papers report data mostly from experiments. Table 3 lists NBS handbooks, bibliographies, computer packages, etc., for mixture data. Figure 2 is taken from (14) and graphically illustrates the available LVE data for mixtures relevant to the gas industry. This same publication contains a quantitative analysis of the ranges and quality of the LVE data and makes

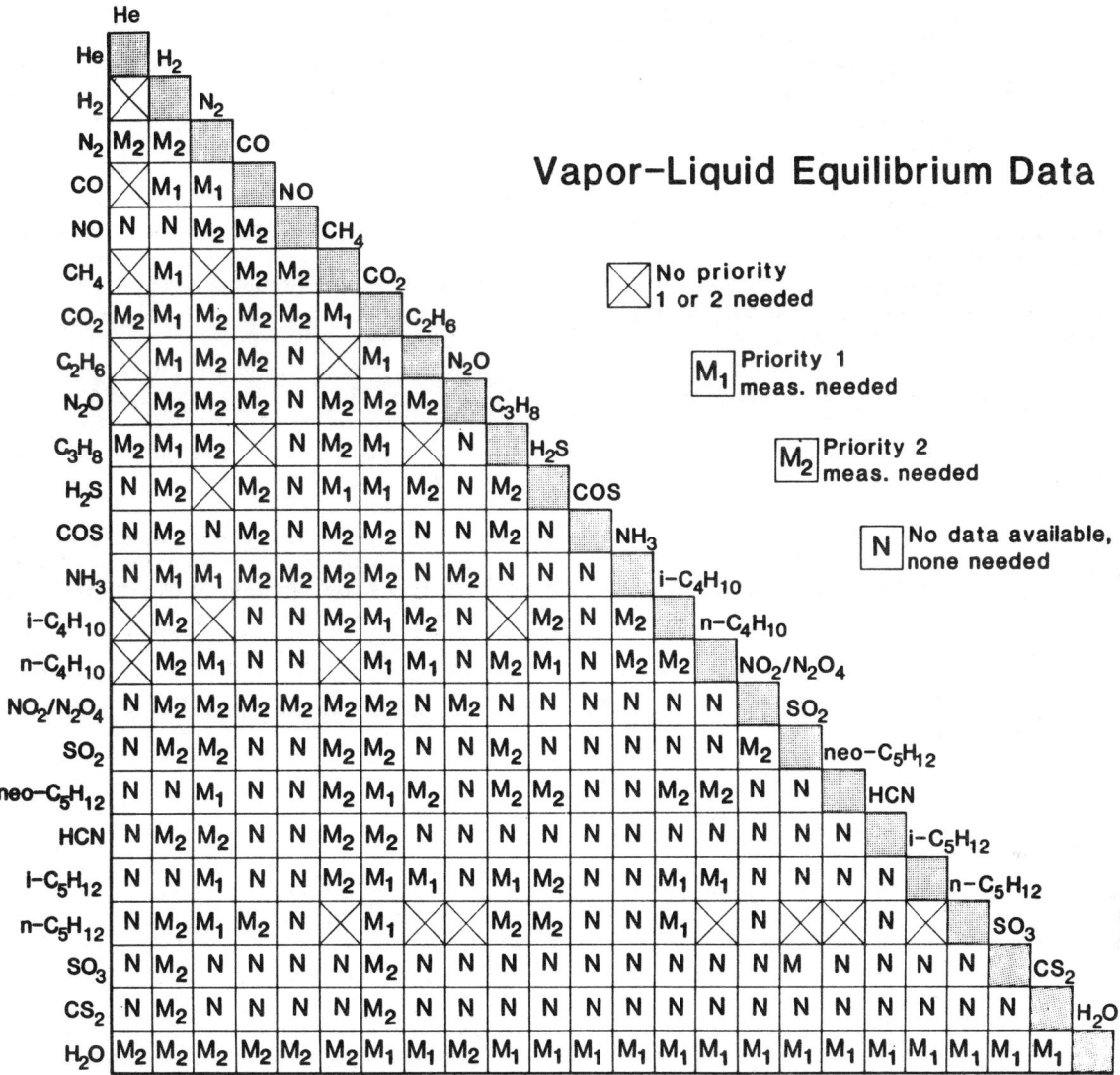

Figure 2. Availability of vapor-liquid equilibrium data and recommendations for measurements.

specific recommendations for needed measurements. The work (14) was performed as part of a larger data project funded by the Gas Research Institute and coordinated by Texas A&M University. The project is significant in that it complements work funded by the NBS-Office of Standard Reference Data [15] and it illustrates the awareness, by an important industrial funding agency, of the vital importance of accurate thermophysical property data.

Solid Property Data. State-of-the-art measurements of the thermal properties of solids are carried out in both the Thermophysics Division and the Chemical Engineering Science Division. In the case of the former, the results are obtained using dynamic techniques for solids at very high temperatures, generally above 2000 K. The work of the latter Division concentrates on the low (cryogenic) and moderate temperature regimes. Table 4 provides a representative collection of these results. You will note that [16] contains both fluid and solid properties.

FUTURE DIRECTIONS

The integration of laboratory measurement on carefully chosen systems with critically evaluated data and theoretical models for the purpose of providing accurate predictive techniques in the form of validated computer programs establishes the modus operandi for future research in thermophysical property data in the NBS-Center for Chemical Engineering. It should be noted that although increased emphasis will be placed on theoretical models, and the development of predictive techniques, a significant part of the total effort will be in laboratory measurements. These measurements will be carefully chosen, but the fact remains that our future demands for thermophysical property data will be in the areas of complex fluids and solids, and the challenge will be to provide the required amount of accurate data to characterize these systems. Some of the particular areas which our future research will encompass include:

1 - High Temperature PVT and Phase Equilibria Measurements
2 - High Temperature Transport Property Measurements; including Non-Newtonian Fluids
3 - Mixture Critical Lines
4 - Fluid Interfaces
5 - Magnetic Suspension Densimetry
6 - Microsecond Measurement Techniques
7 - Mixtures of Polar and Nonpolar Molecules
8 - Polydisperse Fluids
9 - Conformal Solution Theory
10 - Ionic Solutions

In addition, a substantial present and future effort will be placed on the study of the nonlinear behavior of fluids. This latter work is exemplified by the results of an NBS-sponsored meeting held in June 1982 (17).

LITERATURE CITED

1. Baker, O., "The Value of Enthalpy Research to the Petroleum Industry," Proc. 39th Annual Convention Natural Gas Processors Association (1960).

2. Zudkevitch, D., Hydrocarbon Proc. 54, 97 (Mar 1975).

3. Zudkevitch, D. and Gray, R. D., Jr., Adv. Cryogenic Eng. Vol 20, 103 (1975).

4. Zudkevitch, D., "Impact of Thermodynamic and Fluid Properties Data Uncertainties on Design and Economics of Separation Operations," The Encyclopedia of Chemical Processing and Design, J. J. McKetta, Editor.

5. Williams, C. C. and Albright, M. A., Hydrocarbon Proc. 55, 115 (May 1976).

6. Yen, L. C., Frith, J. F. S., Chao, K. C., and Lin, H. M., Chem. Eng. 84, 127 (May 9, 1977).

7. Klein, M., "The Role of Data Accuracy in Applications of Thermophysics: An Introduction and Overview," Nat. Bur. Stand. (U.S.) Spec. Publ. 590, 1-9 (Oct 1980).

8. Khalifa, H. E. and Kestin J., "The Influence of Thermophysical Properties on the Design and Sizing of Geothermal Power Plant Components," Nat. Bur. Stand. (U.S.) Spec. Publ. 590, 19-25 (Oct 1980).

9. Olien, N. A., "Present and Future Sources of Fluid Property Data," Nat. Bur. Stand. (U.S.) Spec. Publ. 590, 11-17 (Oct 1980).

10. Elliot, D. G., Chappelear, P. S. et al., "Thermophysical Properties: Their Effects on Cryogenic Gas Processing," Phase Equilibria and Fluid Properties in the Chemical Industry, ACS Symposium Series, 60, 289-315, T. S. Storvick and S. I. Sandler, Editors, American Chemical Soc. (1977).

11. Miller, E. J., Geist, J. M., "Impact of Recent Developments in Thermodynamics on Chemical Process Design," Presented at the Joint Meeting of the Chemical Industry Engineering Society of China and the AIChE, Beijing (Sep 1982).

12. "National Needs for Critically Evaluated Physical and Chemical Data (the CODAN Report)," National Research Council, National Academy of Sciences, PB 284 646, 69 pages (Jun 1978).

13. Kestin, J., Editor, "Thermophysical Properties for Synthetic Fuels," Division of Engineering, Brown Univ. (Nov 1982).

14. Williamson, F. R. and Olien, N. A., "Compilation and Evaluation of Available Data on Phase Equilibria of Natural and Synthetic Gas Mixtures," Nat. Bur. Stand. (U.S.) NBSIR 83-1692 (1983).

15. Hiza, M. J., Kidnay, A. J., and Miller, R. C., Equilibrium Properties of Fluid Mixtures: A Bibliography of Experimental Data on Selected Fluids," IFI/Plenum, New York, 246 pages (1982).

16 Mann, D. B., Diller, D. E., and Olien, N. A. Editors, <u>LNG Materials and Fluids - A User's Manual of Property Data in Graphic Format</u>, Nat. Bur. Stand. (U.S.) (1977), First Supplement (1979), Second Supplement (1980).

17 Hanley, H. J. M., Editor, <u>Nonlinear Fluid Behavior</u>, Physica <u>118A</u>, Nos. 1-3, North Holland, Amsterdam (1983).

TABLE 1

NBS PURE FLUID PROPERTY PUBLICATIONS

SELECTED PROPERTIES OF HYDROGEN (ENGINEERING DESIGN DATA), by R. D. McCarty, J. Hord, and H. M. Roder. Nat. Bur. Stand. (U.S.) Mono. No. 168, 523 pages (Feb 1981).

THERMODYNAMIC PROPERTIES OF HELIUM 4 FROM 2 TO 1500 K AT PRESSURES TO 10^8 Pa, by R. D. McCarty, J. Phys. Chem. Ref. Data, Vol 2, No. 4, 923-1041 (1973).

THERMOPHYSICAL PROPERTIES OF FLUIDS. 1. ARGON, ETHYLENE, PARAHYDROGEN, NITROGEN, NITROGEN TRIFLUORIDE, AND OXYGEN, by B. A. Younglove, J. Phys, Chem. Ref. Data, Vol 11, Supplement No. 1 (1982).

THERMODYNAMIC PROPERTIES OF COMPRESSED GASEOUS AND LIQUID FLUORINE, by R. Prydz and G. C. Straty, Nat. Bur. Stand. (U.S.), Tech. Note No. 392-Revised, 197 pages (Sep 1973).

THERMODYNAMIC PROPERTIES OF AMMONIA, by L. Haar and J. S. Gallagher, J. Phys. Chem. Ref. Data, Vol 7, No. 3, 635-792 (1978).

THE NBS/NRC STEAM TABLES, by L. Haar, J. S. Gallagher, and G. S. Kell, Hemisphere Press (to be published Oct 1983).

THE THERMOPHYSICAL PROPERTIES OF METHANE, FROM 90 TO 500 K AT PRESSURES TO 700 BAR, by R. D. Goodwin. Nat. Bur. Stand. (U.S.), Tech. Note No. 653, 280 pages (Apr 1974).

THERMOPHYSICAL PROPERTIES OF ETHANE, FROM 90 TO 600 K AT PRESSURES TO 700 BAR, by R. D. Goodwin, H. M. Roder and G. C. Straty. Nat. Bur. Stand. (U.S.), Tech. Note No. 684, 326 pages (Aug 1976).

THERMOPHYSICAL PROPERTIES OF PROPANE FROM 85 TO 700 K AT PRESSURES TO 70 MPa, by R. D. Goodwin and W. M. Haynes. Nat. Bur. Stand. (U.S.) Monogr. No. 170, 249 pages (Apr 1982).

THERMOPHYSICAL PROPERTIES OF NORMAL BUTANE FROM 135 TO 700 K AT PRESSURES TO 70 MPa, by W. M. Haynes and R. D. Goodwin. Nat. Bur. Stand. (U.S.) Monogr. No. 169, 197 pages (Apr 1982).

THERMOPHYSICAL PROPERTIES OF ISOBUTANE FROM 114 TO 700 K AT PRESSURES TO 70 MPa, by R. D. Goodwin and W. M. Haynes. Nat. Bur. Stand. (U.S.) Tech. Note No. 1051, 196 pages (Jan 1982).

A THERMODYNAMIC SURFACE FOR THE CRITICAL REGION OF ETHYLENE, by J. M. H. Levelt Sengers, J. S. Gallagher, F. W. Balfour, and J. V. Sengers. Nat. Bur. Stand. (U.S.) Tech. Note. No. 1165, 82 pages (Oct 1982).

MEASUREMENTS OF THE VISCOSITY OF SATURATED AND COMPRESSED LIQUID METHANE, ETHANE AND PROPANE, by D. E. Diller. Thermophysical Properties, Proc. Symp., 8th, Vol I: Thermophysical Properties of Fluids (Nat. Bur. Stand., Gaithersburg, MD., Jun 15-18, 1981), J. V. Sengers, Editor. American Society of Mechanical Engineers, New York, 219-226 (1982).

THERMAL CONDUCTIVITY OF LIQUID PROPANE, by H. M. Roder and C. A. Nieto de Castro. J. Chem. Eng. Data, Vol 27, No. 1, 12-15 (Jan 1982).

THE VISCOSITY AND THERMAL CONDUCTIVITY COEFFICIENTS FOR DENSE GASEOUS AND LIQUID ARGON, KRYPTON, XENON, NITROGEN, AND OXYGEN, by H. J. M. Hanley, R. D. McCarty, and W. M. Haynes, J. Phys. Chem. Ref. Data, Vol 3, No. 4, 979-1017 (1974).

THE VAPOR PRESSURE, CRITICAL ISOCHORE AND SOME MEASTABLE STATES OF CO_2, by J. M. H. Levelt Sengers and W. T. Chen, J. Chem. Phys., <u>56</u>, 595 (1972).

VISUAL OBSERVATION OF THE CRITICAL TEMPERATURE AND DENSITY: CO_2 AND C_2H_4, by M. R. Moldover, J. Chem. Phys., <u>61</u>, 1766 (1974).

TABLE 2

NBS MIXTURE PROPERTY PUBLICATIONS

LIQUID-VAPOR EQUILIBRIA RESEARCH ON SYSTEMS OF INTEREST IN CRYOGENICS - A SURVEY, by A. J. Kidnay, M. J. Hiza, and R. C. Miller. Cryogenics, Vol 13, No. 10, 575-599 (Oct 1973).

ORTHOBARIC LIQUID DENSITIES AND EXCESS VOLUMES FOR BINARY MIXTURES OF LOW MOLAR-MASS ALKANES AND NITROGEN BETWEEN 105 AND 140 K, by M. J. Hiza, W. M. Haynes, and W. R. Parrish, J. Chem. Thermodyn., Vol. 9, No. 9, 873-896 (1977).

A REVIEW, EVALUATION, AND CORRELATION OF THE PHASE EQUILIBRIA, HEAT OF MIXING AND CHANGE IN VOLUME ON MIXING FOR LIQUID MIXTURES OF METHANE + ETHANE, by M. J. Hiza, R. C. Miller and A. J. Kidnay. J. Phys. Chem. Ref. Data, Vol 8, No. 3, 799-816 (1979).

ORTHOBARIC LIQUID DENSITIES AND EXCESS VOLUMES FOR MULTICOMPONENT MIXTURES OF LOW MOLAR-MASS ALKANES AND NITROGEN BETWEEN 105 AND 125 K, by M. J. Hiza and W. M. Haynes. J. Chem. Thermodyn., Vol. 12, No. 1, 1-10 (Jan 1980).

(P,V,T) OF COMPRESSED AND LIQUEFIED (NITROGEN + METHANE), by G. C. Straty and D. E. Diller. J. Chem. Thermodyn., Vol 12, No. 10, 937-953 (Oct 1980).

A REVIEW, EVALUATION, AND CORRELATION OF THE PHASE EQUILIBRIA, HEAT OF MIXING, AND CHANGE IN VOLUME ON MIXING FOR LIQUID MIXTURES OF METHANE + PROPANE, by R. C. Miller, A. J. Kidnay and M. J. Hiza. J. Phys. Chem. Ref. Data, Vol 9, No. 3, 721-734 (1980).

LIQUID-VAPOR EQUILIBRIA IN BINARY SYSTEMS CONTAINING ^{4}He OR ^{3}He WITH nH_2 OR nD_2, by M. J. Hiza. Fluid Phase Equilib., Vol 6, No. 3-4, 203-227 (Jun 1981).

PREDICTION OF TRANSPORT PROPERTIES. 1. VISCOSITY OF FLUIDS AND MIXTURES, by J. F. Ely and H. J. M. Hanley. Ind. Eng. Chem. Fundam., Vol 20, No. 4, 323-332 (Nov 1981).

MEASUREMENTS OF ORTHOBARIC LIQUID DENSITIES OF MULTICOMPONENT MIXTURES OF LNG COMPONENTS (N_2, CH_4, C_2H_6, C_3H_8, $CH_3CH(CH_3)CH_3$, C_4H_{10}, $CH_3CH(CH_3)C_2H_5$, and C_5H_{12}) BETWEEN 110 AND 130 K, by W. M. Haynes, J. Chem. Thermodyn., Vol 14, No. 7, 603-612 (Jul 1982).

PREDICTION OF TRANSPORT PROPERTIES. 2. THERMAL CONDUCTIVITY OF PURE FLUIDS AND MIXTURES, by J. F. Ely and H. J. M. Hanley. Ind. Eng. Chem. Fundam., Vol 22, No. 1, 90-97 (Feb 1983).

MATHEMATICAL MODELS FOR THE PREDICTION OF LIQUEFIED-NATURAL-GAS DENSITIES, by R. D. McCarty. J. Chem. Thermodyn., Vol 14, No. 9, 837-854 (Sep 1982).

THE SHEAR VISCOSITY COEFFICIENTS OF NITROGEN + METHANE AND METHANE + ETHANE MIXTURES, by D. E. Diller (to be presented at ASME Winter Annual Meeting, Boston, MA, Nov 13-17, 1983).

GRAVITY-INDUCED DENSITY AND CONCENTRATION PROFILES IN BINARY MIXTURES NEAR GAS-LIQUID CRITICAL LINES, by R. F. Chang, J. M. H. Levelt Sengers, T. Doiron, and J. Jones, J. Chem. Phys. (to be published 1983).

ORTHOBARIC LIQUID DENSITIES OF (METHANE + ISOBUTANE) AND (METHANE + NORMAL BUTANE) AT LOW TEMPERATURES, by W. M. Haynes, J. Chem. Thermodyn. (to be published 1983).

CRITICAL POINT MEASUREMENTS ON NEARLY POLYDISPERSE FLUIDS, by G. Morrison and J. Kincaid, AIChE J. (to be published 1983).

TABLE 3

HANDBOOKS, COMPUTER PROGRAMS, BIBLIOGRAPHIES

EQUILIBRIUM PROPERTIES OF FLUID MIXTURES - A Bibliography of Data on Fluids of Cryogenic Interest, by M. J. Hiza, A. J. Kidnay, and R. C. Miller. NSRDS Bibliographic Series, published by IFI/Plenum, New York 166 pages (1975).

LNG MATERIALS AND FLUIDS - A USER'S MANUAL OF PROPERTY DATA IN GRAPHIC FORMAT, D. B. Mann, D. E. Diller, and N. A. Olien Editors. Nat. Bur. Stand. (U.S.), (1977); First Supplement (1979); Second Supplement (1980).

FOUR MATHEMATICAL MODELS FOR THE PREDICTION OF LNG DENSITIES, by R. D. McCarty. Nat. Bur. Stand. (U.S.) Tech. Note No. 1030, 84 pages (Dec 1980). (Companion to R. D. McCarty reference listed in Table 2).

INTERACTIVE FORTRAN PROGRAM TO CALCULATE THERMOPHYSICAL PROPERTIES OF SIX FLUIDS, by B. A. Younglove. Nat. Bur. Stand. (U.S.) Tech. Note No. 1048, 56 pages (Jun 1982). (Companion to B. A. Younglove reference listed in Table 1).

EQUILIBRIUM PROPERTIES OF FLUID MIXTURES - 2. A BIBLIOGRAPHY OF EXPERIMENTAL DATA ON SELECTED FLUIDS, by M. J. Hiza, A. J. Kidnay, and R. C. Miller. NSRDS Bibliographic Series, Published by IFI/Plenum Press, New York, 253 pages (1982).

A COMPUTER PROGRAM FOR THE PREDICTION OF VISCOSITY AND THERMAL CONDUCTIVITY IN HYDROCARBON MIXTURES, by J. F. Ely and H. J. M. Hanley. Nat. Bur. Stand. (U.S.) Tech. Note No. 1039, 80 pages (Apr 1983). (Companion to J. F. Ely references listed in Table 2).

COMPILATION AND EVALUATION OF AVAILABLE DATA ON PHASE EQUILIBRIA OF NATURAL AND SYNTHETIC GAS MIXTURES, by F. R. Williamson and N. A. Olien. Nat. Bur. Stand. (U.S.) NBSIR 83-1692 (to be published Jul 1983).

TABLE 4

NBS SOLID PROPERTY PUBLICATIONS

THERMAL CONDUCTIVITY AND ELECTRICAL RESISTIVITY STANDARD REFERENCE MATERIALS: AUSTENITIC STAINLESS STEEL, SRM'S 735 AND 798, FROM 4 TO 1200 K, by J. G. Hust and P. J. Giarratano. Nat. Bur. Stand. (U.S.), Spec. Publ. No. 260-46, 42 pages (Mar 1975).

THERMAL CONDUCTIVITY AND ELECTRICAL RESISTIVITY STANDARD REFERENCE MATERIALS: ELECTROLYTIC IRON, SRM'S 734 AND 797 FROM 4 TO 1000 K, by J. G. Hust and P. J. Giarratano. Nat. Bur. Stand. (U.S.), Spec. Publ. No. 260-40, 41 pages (Jun 1975).

THERMAL CONDUCTIVITY AND ELECTRICAL RESISTIVITY STANDARD REFERENCE MATERIALS: TUNGSTEN, SRM'S 730 AND 799 FROM 4 TO 3000 K, by J. G. Hust and P. J. Giarratano. Nat. Bur. Stand. (U.S.), Spec. Publ. No. 260-52, 47 pages (Sep 1975).

A SUBSECOND PULSE HEATING TECHNIQUE FOR THE STUDY OF SOLID-SOLID PHASE TRANSFORMATION AT HIGH TEMPERATURES: APPLICATION TO IRON, by A. Cezairliyan and J. L. McClure, High Temperature Science, Vol 7, pages 189, (1975).

ENTHALPY AND HEAT CAPACITY STANDARD REFERENCE MATERIAL - MOLYBDENUM SRM 781, FROM 273 TO 2800 K, by D. A. Ditmars, A. Cezairliyan, S. Ishihara, and T. B. Douglas. Nat. Bur. Stand. (U.S.), Spec. Publ. No. 260-55 (1977).

ADVANCES IN MEASUREMENTS OF THERMOPHYSICAL PROPERTIES BY DYNAMIC TECHNIQUES, by A. Cezairliyan, High Temperature-High Pressures, Vol 11, 9 (1979).

A TRANSIENT (SUBSECOND) TECHNIQUE FOR MEASURING HEAT OF FUSION OF METALS, by A. Cezairliyan, Int. J. Thermophysics, Vol 1, 83 (1980).

SPECIFIC HEAT CAPACITY AND ELECTRICAL RESISTIVITY OF A CARBON-CARBON COMPOSITE IN THE RANGE 1500-3000 K BY A PULSE HEATING METHOD, by A. Cezairliyan and A. P. Miiller, Int. J. Thermophysics, Vol 1, 317 (1980).

MEASUREMENT OF EFFECTIVE THERMAL CONDUCTIVITY OF A GLASS FIBERBOARD STANDARD REFERENCE MATERIAL, by D. R. Smith and J. G. Hust. Cryogenics Vol 21, No. 7, 408-410 (Jul 1981).

MEASUREMENT OF EFFECTIVE THERMAL CONDUCTIVITY OF GLASS FIBER BLANKET STANDARD REFERENCE MATERIAL, by D. R. Smith, J. G. Hust, and L. J. Van Poolen. Cryogenics Vol 21, No. 8, 460-462 (Aug 1981).

LOW-TEMPERATURE PROPERTIES OF EXPANDED POLYURETHANE AND POLYSTYRENE, by L. L. Sparks. Thermal Insulation Performance, Proc. Symp. (Tampa, FL., Oct 23-25, 1978), D. L. McElroy and R. P. Tye, Editors. American Society for Testing and Materials, Philadelphia, PA, Spec. Tech. Publ. No. ASTM STP 718, 431-452 (1981).

THERMAL CONDUCTIVITY OF A POLYURETHANE FOAM FROM 95 K TO 340 K, by L. L. Sparks. Nat. Bur. Stand. (U.S.) NBSIR 82-1664 22 pages (Mar 1982).

REVIEW OF NEEDS FOR THERMOPHYSICAL PROPERTY DATA ON SOLID FEEDSTOCKS. 1. COAL, by J. E. Callanan. Nat. Bur. Stand. (U.S.) NBSIR 82-1666, 25 p. (May 1982).

RADIANCE TEMPERATURE OF METALS AT THEIR MELTING POINTS AS POSSIBLE HIGH TEMPERATURE SECONDARY REFERENCE POINTS, by A. Cezairliyan, A. P. Miiller, F. Righini, and A. Rosso, Temperature: Its Measurement and Control in Science and Industry, Vol 5, J. F. Schooley, Ed., American Institute of Physics, New York, 377 (1982).

EFFECTIVE THERMAL CONDUCTIVITY OF GLASS-FIBER BOARD AND BLANKET STANDARD REFERENCE MATERIALS, by D. R. Smith and J. G. Hust, 17th International Thermal Conductivity Conference, June 15-18, 1981, Gaithersburg, MD, Plenum Press, New York and London 793 (1982).

THERMAL CONDUCTIVITY OF CONCRETE MORTAR, by L. L. Sparks, 17th International Thermal Conductivity Conference, June 15-18, 1981, Gaithersburg, MD, J. G. Hust Ed., Plenum Press, New York, 655-663 (1983).

THERMODYNAMIC AND THERMOCHEMICAL STUDIES AT THE BARTLESVILLE ENERGY TECHNOLOGY CENTER

B.E. Gammon ■ U.S. Department of Energy, Bartlesville Energy Technology Center, Bartlesville, OK 74005

The Bartlesville Energy Technology Center/National Institute for Petroleum and Energy Reserach (BETC/NIPER) provides thermodynamic information of utility in the extraction, processing and utilization of organic fossil substances. Current and recent thermodynamic studies will be discussed. The studies for pure compounds employ combustion calorimetry, condensed and vapor-phase heat-capacity calorimetry, PVT, speed-of-sound, vapor-pressure, and spectroscopy measurements with derived statistical thermodynamic functions. Vapro-liquid-equilibria on coal liquids and thermodynamic behavior for enhanced oil recovery are also determined.

INTRODUCTION

The Bartlesville Energy Technology Center/National Institute for Petroleum and Energy Research (BETC/NIPER) [1][2] currently of the U.S. Department of Energy has maintained a set of laboratories devoted to providing thermodynamic information of utility in the extraction, processing and utilization of organic fossil materials which include petroleum, coal and oil shale.

As it has for many years, this laboratory provides data on pure organic compounds for predicting chemical equilibria and thus the thermodynamic limits or goals on yields and conversions for many processes. For such predictions one must have the change in the Gibbs energy, ΔG, at any temperature and pressure of interest to calculate the equilibrium constant, $K°$, in equation [1],

$$\Delta G° = -RT \ln K°. \qquad [1]$$

For this purpose, studies are made on "key" compounds from pertinent chemical families to provide essential data for the prediction of the thermochemical properties of related compounds for which measurements have not been made. The individual results obtained for use in deriving ΔG for the ideal gaseous state include the enthalpies of combustion, heat capacities, entropies and enthalpies along the saturation lines, vapor pressures, enthalpies of vaporization, vapor heat capacities, and spectroscopically derived thermodynamic functions. Separately each of these pieces of data is of utility to the engineer. When taken collectively with imposed thermodynamic consistency, these data provide a very reliable source for the prediction of thermodynamic properties to very high temperatures where many processes are operated but where the compounds are too unstable for measurements to be made on a single species.

In addition, these laboratories have provided results of high accuracy and over very broad ranges of the state surface of compressed fluids through PVT and speed of sound measurements.

In recent years, extensive studies have been made at BETC on materials taken from coal conversion process streams and on synthetic multicomponent mixtures related to such processes. These studies have included hydrogen solubility with related vapor-liquid equilibria measurements, enthalpies of combustion, and heat capacities by differential scanning calorimetry.

Thermodynamic investigations are being made at BETC to gain a basic knowledge associated with enhanced oil recovery. These

---

B.E. Gammon is Chief, Thermodynamics Research Branch, Bartlesville Energy Technology Center, now National Institute for Petroleum and Energy Research.

Numbers in parentheses refer to items in the list of references at the end of this report.

studies employ solution calorimetry, densimetry, adsorption calorimetry, both small-angle X-ray and visible light scattering, ultracentrifugation, and isopiestic distillation to delineate the basic mechanisms involved in the production of oil through the addition to the reservoirs of selected chemical surfactants, polymers, etc., with suitable driving fluids, particularly in the aqueous phase.

In the discussion that follows, emphasis will be placed on organic substance studies for which the laboratories at Bartlesville have a well established role and discipline for providing an appreciable volume of high quality data of interest to this symposium. The laboratory's roles in other areas are being vigorously pursued but will be only briefly described.

EXPERIMENTAL FACILITIES

The experimental disciplines used for providing the thermodynamic data at Bartlesville are listed in the appendix with brief descriptions of the equipment employed and the products provided.

CHEMICAL THERMODYNAMIC PROPERTIES OF PURE COMPOUNDS

$\Delta_f H$, C_p, S, P_s, $\Delta_v H$, and Ideal Gas Properties

Limitations exist in the range of applicability of experimental methods for the measurements of thermodynamic properties of pure organic compounds owing to increased instability and/or reactivity as the temperature is increased. For many such compounds this barrier is reached in the range of 250 to 300 C. This is a barrier to pure compound studies but conversely is a threshold where useful conversion reactions can occur and catalytic requirements begin to diminish. With these provisions, the rationale and current limitations for measurements at BETC/NIPER are delineated, then current compound studies are presented and discussed. Sources of thermochemical data on organic compounds are discussed in appendix B.

Experimental Techniques at BETC/NIPER

In the pure compound studies, the enthalpy of formation $\Delta_f H$, is obtained from measurements of the enthalpy of combustion of the substance in the condensed state at 298.15 K. Measurements of the heat capacity and the enthalpies of transition are made at temperatures from near 0 K to well into the liquid phase, where feasible, to provide values of the entropy of formation, $\Delta_f S$, from the application of the third law of thermodynamics. From these, one can obtain the Gibbs energy of formation in the condensed state, $\Delta_f G = \Delta_f H - T\Delta_f S$, at any temperature covered by the heat capacity and enthalpy measurements.

For many processes, the properties of most interest are in the gaseous phase, and correlation of properties $\Delta_f G$, $\Delta_f H$, and $\Delta_f S$ within compound families can be most feasibly done in the ideal gaseous state where contributions from intermolecular interaction are absent and thus can be treated as a separate state property problem. To provide the link to the gaseous phase, vapor pressures are determined to give the Gibbs energy and entropy of compression from the saturation line to the ideal gaseous state at one atmosphere. The enthalpy of vaporization is also determined via the Clapeyron equation. The above methods are applied to compounds whose vapor pressures reach at least 0.001 bar at 573 K.

For lower molecular weight materials which have vapor pressures of 2 bars below 438 K, enthalpies of vaporization and vapor heat capacities are determined by direct measurement. The vapor-heat-capacity results provide important checks and a source for semiempirical adjustment of thermodynamic functions derived from spectroscopic measurements via statistical thermodynamics. Also the vapor-heat-capacity data provide very reliable values of the second virial coefficients which are not perturbed by adsorption effects found in PVT measurements in this pressure range.

Properties derived from spectroscopic measurements and augmented by direct experimental measurements provide results at high temperatures where an organic substance can only exist in (quasi) equilibrium with other substances. Presently the derivation of accurate values of the thermodynamic properties from molecular statistical mechanical calculations is limited to those compounds whose vapor pressures reach 1 bar before they decompose. This limitation arises because suitable vapor-phase Raman spectra needed for the determination of low frequency molecular vibrations can only be obtained at molecular densities near 1 bar.

Current and Recent Studies at BETC

In recent years studies have been and continue to be directed toward compounds derived from coal, shale oil, and heavy

petroleum to answer questions concerning conversion efficiencies and concerning strategies for the removal of deleterious materials known to form gums, to cause air pollution problems and to be health hazards. The compound families include:
(1) alkylnaphthalenes, alkylindans and related compounds.
(2) polycyclic aromatic hydrocarbons and their hydrogenation products.
(3) polycyclic nitrogen compounds and their hydrogenation products.
(4) polycyclic oxygen compounds and their hydrogenation products.
(5) low molecular weight nitrogen compounds and their hydrogenation products.

Though most of the work in recent years has been on high-molecular-weight substances, studies on low-molecular-weight nitrogen compounds are being made to fill needs for the utilization of shale oil. Studies have been completed, and results are being prepared for publication on the compounds listed in Table 1. Studies have also been conducted on bridged-ring compounds which have high energies per unit volume and thus are candidates for fuels in vehicles with limited space for storage. The majority of the compounds in this work were synthesized and purified by Professor E. J. Eisenbraun's group at Oklahoma State University, a natural extension of their earlier work on hydrocarbons for the American Petroleum Institute on Project 58A.

Significant progress has been made in the calculation of thermodynamic properties from data derived from spectra. Studies (2-4) have shown that previously reported differences between experimental and calculated entropies were often a result of using liquid spectra rather than vapor spectra to assign fundamental vibrational frequencies and not from postulated barriers to rotation and other empiricisms. Also, progress has been made in calculating properties of substances which undergo inversions such as those found in nitrogen compounds and in five-membered rings (5).

PROPERTIES OF COMPRESSED FLUIDS

With the exception of combustion calorimetry, all of the studies at low pressure described in the previous section help define state behavior which is important in many designs for extraction, conversion, and utilization processes for fossil substances. Large regions of state surfaces are investigated at higher pressures by the methods described in this section.

Table 1. Compounds with Data to be Published

Compound	Properties Available[1]		
	Δ_fH	S	P_s
4-Methylphenanthrene	U	U	U
α,α'-Bibenzyl	A	U	P
2,2'-Dimethylbiphenyl	U	U	U
1,2,3,4,5,6,7,8-Octahydroanthracene	U	P	U
Phenyl-2-tolylmethane	U	U	U
2,3-Dimethylnaphthalene	P	U	P
1-Isopropyl-8-methyl-napthalene	U		
1-Ethyl-8-methylnaphthalene	U		
2-Ethyl-6-methylnaphthalene	U		
1,1',2,2',3,3',4,4'-Octahydro-2,2'-binaphthyl	U		
1-Isopropyl-6-methylindan	U		
1,6-Dimethylindan	U		
2,2'-Biindanyl	U		
1,4-Dimethylbenzene	A	U	P
Norbornane	A	U	P
1,1'-Bicyclopentyl	P	U	
1-Methyl-1-ethylcyclopentane	P	U	P
1-Pentene	P	U	P
1-Nonene		U	A
1-Hexadecene	A	U	P
cis-2-Hexene	A	U	A
Phenanthridine	A	U	
Acridine	U	U	A
9-Methylcarbazole	P	U	
Isoquinoline	U	U	U
Quinoline	U	U	U
2,6-Dimethylpyridine	A	U	A
3,5-Dimethylpyridine	A	U	A
2,4-Dimethylpyridine	A	U	A
4-Methylpyridine	P	U	P
2-Methylpiperidine	P	U	P
Piperidine	P	U	P
1-Methylpyrrole	P	U	P
2,5-Dimethylpyrrole	P	U	P
3-Methylpyrrolidine	U	U	U
2,3-Benzofuran	U	U	A
Tetrahydrofuran	A	U	P
Oxetane	A	U	P
2-Ethyloxirane	U	U	P
trans-2,3-Dimethyloxirane	U	U	P

[1]The heading symbols, Δ_fH, S & P_s, designate combustion calorimetric, low-temperature heat capacity, and vapor pressure data, respectively. The letters in the table, U and P, designate unpublished data and published data determined at BETC/NIPER, and A designates that data is available in the literature.

Two inactive pieces of equipment have been used at BETC/NIPER for the measurement of state properties of pure and binary mixtures of compressed fluids. A volumetric mercury compression apparatus provides PVT data at temperatures from 238 to 623 K and to pressures of 450 bar. Virial coefficients and the thermodynamic functions, enthalpy, entropy and Gibbs energy are usually derived from the data. This apparatus has been idle but is being reactivated; it was most recently used in studies on propane (6) and ethylene (7,8).

The other inactive apparatus is an ultrasonic interferometer which provides the speed of sound in fluids at temperatures from 98 to 573 K and at pressures from 0.001 to 300 bars. It has been used to derive virial coefficients, constant volume heat capacities, and other state properties. In recent work, measurements were made in ethylene (9) at temperatures from 104 to 298 K as part of a comprehensive program coordinated by the National Standard Reference Data System of the National Bureau of Standards. This apparatus has also been used to provide data on methane (10) and helium (11).

In recent years an apparatus was developed for the determination of hydrogen solubility, and vapor-liquid equilibria of multicomponent fluids associated with coal conversion processes. It employs a stirred autoclave suitable for hydrogen and operates at temperatures to 800 F and to pressures of 4000 psia. It has been used to determine the solubility of hydrogen in a synthetic recycle solvent comprised of tetralin, 2-methylnaphthalene, p-cresol, and 4-picoline, and also used in middle and heavy distillates from the Solvent Refined Coal (SRC) II process (12, 13). The apparatus has been used to determine hydrogen solubilities, phase densities, and vapor-liquid compositions on:
1) three blends of H-coal process liquids with water and a nine-component light gas mixture of H_2, N_2, CO_2, H_2S, C_1-C_4, H_2O, and NH_3 (14,15),
2) systems of tetralin, 2-methylnaphthalene, p-cresol, 4-picoline, water, ammonia and hydrogen (16), and
3) 1-naphthol, quinoline, tetralin, water, ammonia and hydrogen (17).

These latter studies required the full analytical support of BETC's Characterization Branch with their variety of chromatographic, mass spectroscopic and electromagnetic spectroscopic methods to provide the requisite analysis. As an integral part of this work, heat capacities and enthalpies of combustion were determined on chars and many well characterized fluids obtained from coal conversion processes (18).

STUDIES FOR ENHANCED OIL RECOVERY

For several years, the primary mission of BETC/NIPER has been to conduct research on enhanced oil recovery processes. It is estimated that enhanced oil recovery processes which include chemical flooding, gas miscible flooding, and thermal methods can be used to produce an additional 20 percent or more of the oil remaining in reservoirs after primary and secondary production methods have been terminated.

For the development of surfactant flooding techniques, many laboratory studies at BETC/NIPER are concerned with understanding the properties and interactions of the surfactants, cosurfactants, polymers, brines, hydrocarbons, and the rock matrices of the reservoirs. In the thermodynamic studies at BETC/NIPER, part of the work has been devoted to understanding the behavior of surfactant solutions in the fluid phase and to understanding the mechanisms of adsorption of components from solutions on to solids. Microcalorimeters and highly sensitive densimeters are used to determine densities, heat capacities, and enthalpies of dilution of solutions; enthalpies of immersion of solids into solutions; and enthalpies of preferential adsorption of components from solution on to solids. Recent studies were completed to show a comparison of observed heats of adsorption of liquid mixtures on solid surfaces (19) and the effect of the chain length of cosurfactants on adsorption (20). Also a comprehensive model was recently developed for the thermodynamic properties of surfactant solutions (21,22).

A second part of the present work is concerned with how colloidal properties of micelles and oils are influenced by additives and induced physical changes within reservoirs. Ultracentrifugation augmented with isopiestic distillation experiments, light scattering experiments, and small-angle X-ray scattering studies are made in these investigations. Work was recently completed to show effects of co-ions on micelle aggregation numbers (23) and to show general methods of calculation of the polarization factor for multiple coherent scattering of unpolarized and plane-polarized X-rays (24).

DISCUSSION

For those interested in published studies, a list of publications with compound cross references is maintained on all of the thermodynamic work done at Bartlesville since the group was established. Copies are available for distribution.

The Bartlesville Energy Technology Center is currently undergoing a transition in its operation (1); however thermodynamic investigations are expected to continue to play a significant role at the research center after it becomes the National Institute for Petroleum and Energy Research (NIPER) under the direction of the IIT Research Institute (IITRI). The U.S. Department of Energy has made a commitment to support many of the current thermodynamic research programs, and IITRI has made a commitment to solicit new thermodynamic research projects from organizations funded by the private sector and other government agencies.

It is expected that some of the currently severe constraints in staffing will be relieved to permit these new projects to be pursued. In addition many of the instruments are being automated to thereby minimize manpower requirements. Replicate equipment is being constructed for experimental results that are slowed by relaxation times in the properties being measured. New equipment is being developed for faster measurement of data whose acquisition rate is presently limited by the previous choice of procedures and equipment design.

For studies of high-molecular-weight substances, newly fabricated equipment is being put into operation, and other equipment is being developed to extend the temperature range of the measurements. Measurement techiques which require short residence times at high temperatures are being developed.

There has been a renewed interest in the utility and need for thermodynamic data in recent years (25-31), and it is expected that efforts at Bartlesville will be made to provide significant contributions of such data in carefully disciplined studies that will ensure its reliability and lasting value for applications to many future questions that can be best answered by use of the predictions so admirably made with thermodynamics.

ACKNOWLEDGEMENT

The dedicated thermodynamicists and support personnel who have toiled in building the thermodynamics program at Bartlesville over the years of its existence are to be commended. The faithful support of the Bureau of Mines, the Energy Research and Development Administration, and the Department of Energy has been instrumental in the continuity of efforts which are essential for the measurement of reliable, extensive sets of thermodynamic data. Acknowledgement also should be made of the efforts of several DOE research administrators who have understood the value of thermodynamic and thermochemical research and who have helped to augment its funding in recent years. Prominent among these administrators were F. Dee Stevenson of the Office of Basic Energy Sciences, Office of Energy Research, Robert Roberts, and Richard J. Bergemann of the Division of Fossil Energy, Washington DOE Headquarters, and Gilbert V. McGurl of the Pittsburgh Energy Technology Center.

June Forbes and Sharon Johnson are thanked for the final preparation of this manuscript, Contribution No. 265 from the thermodynamics research laboratory at the Bartlesville Energy Technology Center, Department of Energy.

APPENDIX A

THERMODYNAMICS RESEARCH DISCIPLINES AT THE BARTLESVILLE ENERGY TECHNOLOGY CENTER

1. Combustion Calorimetry
   Equipment: Rotating-bomb calorimeters (32-34) equipped with quartz thermometry and automated data collection.
   Products: Enthalpy of combustion in the condensed state of pure compounds with derived enthalpy of formation. Enthalpies of combustion of complex mixtures such as coals, chars, coal-derived fluids, finished fuels, etc.

2. Heat Capacity Calorimetry
   Equipment: Adiabatic heat capacity calorimeters with four fully instrumented cryostats (35-37) and with automatic data collection nearly developed.
   Products: Heat capacities, enthalpies of transition, purity determined in the range from 4 K to 573 K and 0 to 2 bar. Entropy and Gibbs energy derived for pure compounds.
   Equipment: Differential scanning calorimeter.
   Products: Constant volume heat capacity of pure compounds, coal, char, and coal-derived fluids from 100 K to 1000 K and at pressures to 300 bar. Purity determined.

3. Vapor-Pressure
   Equipment: Ebulliometers (38-40) with automatic data collection being developed (41).
   Products: Vapor pressures from 0.09 to 2.7 bar with derived enthalpies of vaporization.
   Equipment: Inclined-piston gage (42-44) with automatic data collection being developed (41).
   Products: Vapor pressure from 0.1 mbar to 50 mbar with derived enthalpies of vaporization.

4. Liquid Densimetry
   Equipment: Vibrating tube densimeter being developed.
   Products: Densities of liquids along the vapor-liquid saturation line at temperatures from 263 to 423 K +.

5. Molecular Spectroscopy and Statistical Thermodynamics
   Equipment: Raman and far infrared spectrometers with modifications for heated cells and computer-controlled data collection.
   Products: Molecular spectra for the vapor state with sample temperatures from

298 to 523 K, and with derived statistical thermodynamic properties for the ideal gas at temperatures to 1500 K.
6. Vapor Heat Capacity Calorimetry
   Equipment: Vapor flow calorimeter (45-47) with automatic data collection being developed.
   Products: Enthalpies of vaporization from 288 to 438 K and vapor heat capacities from 298 to 523 K with pressures from 0.125 to 2 bars. Derived second virial coefficients, and ideal gas heat capacities.
7. Pressure-Volume-Temperature Research
   Equipment: Volumetric mercury compression apparatus (48-50).
   Products: P-V-T properties of liquids and gases from 238 K to 623 K with pressures to 450 bars. Derived virial coefficients, and thermodynamic functions, enthalpy, entropy, and Gibbs energy.
8. Speed of Sound in Compressed Fluids
   Equipment: Double-crystal, ultrasonic interferometer (10,51).
   Products: Speed of sound in compressed gases and liquids, at temperatures from 98 K to 573 K with pressures from 0.001 bar to 300 bars. Derived virial coefficients, constant volume heat capacity and other state properties.
9. Vapor-Liquid Equilibria
   Equipment: Stirred autoclave suitable for hydrogen (13) with access to a wide variety of chromatographic and of mass and electromagnetic spectroscopic instruments for analysis.
   Products: Hydrogen solubilities, K values, and densities of phases on multicomponent systems at pressures to 4000 psia and temperatures to 800 F.
10. Solution and Adsorption Calorimetry and Densimetry of Solutions
    Equipment: Microsolution calorimeters, microadsorption calorimeters and solution flow calorimeter-densimeter with automatic data collection and reduction (52).
    Products: Heat capacities, enthalpies of solution, dilution, micellization and adsorption, at ambient pressures and at 15 to 55 C also densities partial molal values deduced.
11. Ultracentrifugation
    Equipment: Ultracentrifuge with automated data collection and analysis and isopiestic distillation equipment.
    Products: Structural information concerning colloids and micelles in petroleum fluids at temperatures from -20 C to 55 C.
12. Light Scattering
    Equipment: Brice-Phoenix intensity light scattering photometer.
    Products: Structural information concerning colloids and micelles in petroleum fluids at temperatures from -20 C to 55 C.
13. X-ray Scattering
    Equipment: Automated Bonse-Hart small-angle X-ray scattering unit and associated equipment.
    Products: Structural information concerning colloids and micelles in petroleum fluids.

APPENDIX B

CONVENIENT SOURCES OF THERMOCHEMICAL DATA

This section gives some convenient sources for current and for critically evaluated and carefully documented thermochemical data on organic compounds. Much of this information can be gleaned from the Thermochemical Bulletin which is described by its editor in these proceedings.

An outstanding set of review articles was recently published on lectures presented at the Twelfth International Conference on Chemical Thermodynamics held under the auspices of the IUPAC Commission on Thermodynamics at the University College, London, U.K., Sept. 6-10, 1982. The reviews by Mansson (53), Suga (54), and Knobler (55) are particularly germane to thermochemical data.

The Thermodynamics Research Center at Texas A&M University maintains and distributes loose-leaf tables (56) on properties of hydrocarbons and related compounds; they also provide critically assessed data in special publications such as their API Monographs (57). Stull, Westrum, and Sinke (58) and Karapet'yants and Karapet'yants (59) made extensive compilations on the chemical thermodynamic properties of organic compounds. Data from the foregoing may be used for the calculation of equilibrium constants for reactions of organic compounds.

Cox and Pilcher (60) compiled standard enthalpies of formation, of combustion, and of vaporization on organic and organometallic compounds at 298.15 K. Pedley and Rylance (61) subsequently extended this compilation and used least-squares adjustments of available data to ensure thermodynamic consistency of data in reaction networks.

Boublik, Fried, and Hala (62) made an extensive compilation of vapor pressures of pure substances. Other vapor pressure compilations of note can be found in references (63) and (64), and a compilation is expected in DeChema Chemistry Data Series (65). Dymond and Smith (66) published a second virial coefficient compilation that has proven to be useful in the evaluation of thermochemical data.

Spectra for use in molecular-statistical-mechanical calculation of thermodynamic properties are compiled at the Molecular Spectra Center at the National Bureau of Standards in Washington, D.C., at the Thermodynamics Research Center at Texas A&M University, and at Sadtler Research Laboratories in Philadelphia, Pa.

The foregoing is not a comprehensive listing, but it should give the reader a start on the retrieval of carefully evaluated and compiled data employed in thermochemistry.

LITERATURE CITED

1. On October 1, 1983, the direct U.S. Government operation of the Bartlesville Energy Technology Center (BETC) ceased. At that time the direct operation of the Center was assumed by the IIT Research Institute, and the Center was renamed the National Institute for Petroleum and Energy Research (NIPER), with a charter to conduct research for the private sector as well as U.S. Government and State supported agencies.
2. J. A. Draeger and D. W. Scott, J. Chem. Phys. 74, 4748 (1981).
3. J. A. Draeger and D. W. Scott, J. Chem. Phys. 75, 2016 (1981).
4. J. A. Draeger and D. W. Scott, J. Chem. Thermodynamics 14, 991 (1982).
5. J. A. Draeger, R. H. Harrison, and W. D. Good, J. Chem. Thermodynamics 15, 367 (1983).
6. R.H.P. Thomas and R. H. Harrison, J. Chem. Eng. Data 27, 1 (1982).
7. R. H. Harrison and D. R. Douslin, J. Chem. Eng. Data 22, 24 (1977).
8. D. R. Douslin and R. H. Harrison, J. Chem. Thermodynamics 8, 301 (1976).
9. B. E. Gammon, to be published.
10. B. E. Gammon and D. R. Douslin, J. Chem. Phys. 64, 203 (1976).
11. B. E. Gammon, J. Chem. Phys. 64, 2556 (1976).
12. R. H. Harrison, oral presentation, American Chemical Society, Las Vegas, Nev., August 1980.
13. R. H. Harrison, S. E. Scheppele, G. P. Sturm, Jr., and P. L. Grizzle, to be published.
14. R. H. Harrison, J. W. Vogh, P. L. Grizzle, and J. S. Thomson. Div. of Fuel Chemistry, American Chemical Society Preprints 27, No. 3-4, 98-103, Kansas City, Mo., Sept. 12-17, 1982.
15. R. H. Harrison, J. W. Vogh, P. L. Grizzle, and J. S. Thomson, to be published.
16. R. H. Harrison, J. W. Vogh, and J. B. Green, to be published.
17. D. W. Arnold, R. H. Harrison, J. W. Vogh, J. B. Green, and M. K. Gupta, to be published.
18. N. K. Smith, S. H. Lee-Bechtold, and W. D. Good, "Thermodynamic Properties of Materials Derived from Coal Liquefaction," DOE/BETC/TPR-79/2, 1980, 25 pp, available through the National Technical Information Service, U.S. Department of Commerce, Springfield, VA 22161.
19. G. W. Woodbury, Jr., and L. A. Noll, Colloids and Surfaces, 8, 1-15 (1983).
20. L. A. Noll, G. W. Woodbury, Jr., and T. E. Burchfield, submitted for publication.
21. T. E. Burchfield and E. M. Woolley, submitted for publication.
22. E. M. Woolley and T. E. Burchfield, submitted for publication.
23. D. A. Doughty, J. Phys. Chem. 87, No. 25, 5286-5290 (1983).
24. C. W. Dwiggins, Acta Crystallographica, A39, 773-777 (1983).
25. C. E. Holly, C. Vanderzee, and E. F. Westrum, Jr., Bulletin of Chemical Thermodynamics, Vol. 21, 476 (1978).
26. N. A. Olien, Proceedings of the Winter Annual Meeting, American Society of Mechanical Engineers, New York, N.Y., Dec. 2-7, 1979.
27. S. C. Stinson, Chem. Eng. News, Jan. 3, 1983, p. 34.
28. D. Garvin, V. B. Parker, and D. D. Wagman, Chemtech, November 1982, p. 691.
29. J. Kestin, "Thermophysical Properties for Synthetic Fuels," Report of Working Group on Thermophysical Properties of Synthetic and Related Fuels, under contract with the Office of Basic Energy Sciences, U.S. Department of Energy, November 1982.
30. Recon Systems, Inc., "Fundamental Data Needs for Coal Conversion Technology," October 1977, revised January 1981. DOE/TID-28152, available through the National Technical Information Service, U. S. Department of Commerce, Springfield, VA 22161.
31. D. W. Brinkman, M. J. Reilly, editors, "Design Properties of Coal Liquids: Edited Workshop Proceedings," August 1981, U.S. Department of Energy CONF-810381, available through the National Technical Information Service, U.S. De-

partment of Commerce, Springfield, VA 22161.
32. W. D. Good, D. R. Douslin, D. W. Scott, A. George, J. L. Lacina, J. P. Dawson, and G. Waddington, J. Phys. Chem. $\underline{63}$, 1133 (1959).
33. W. D. Good, D. W. Scott, and G. Waddington, J. Phys. Chem. $\underline{60}$, 1080 (1956).
34. W. N. Hubbard, C. Katz, and G. Waddington, J. Phys. Chem. $\underline{58}$, 142 (1954).
35. H. M. Huffman, Chem. Rev. $\underline{40}$, 1 (1947).
36. R. A. Ruehrwein and H. M. Huffman, J. Amer. Chem. Soc. $\underline{65}$, 1620 (1943).
37. B. E. Gammon, et al, unpublished on subsequent versions based on designs from refs. 35 and 36.
38. A. G. Osborn and D. R. Douslin, J. Chem. Eng. Data $\underline{11}$, 502 (1966).
39. W. Swietoslawski," Ebulliometric Measurements," Reinhold Pub. Corp., New York, 1945, p. 11.
40. G. Waddington, J. W. Knowlton, D. W. Scott, G. D. Oliver, S. S. Todd, W. N. Hubbard, J. C. Smith, and H. M. Huffman, J. Amer. Chem. Soc. $\underline{71}$, 797 (1949).
41. B. E. Gammon, et al, unpublished.
42. D. R. Douslin and J. P. McCullough, U.S. BuMines RI 6149, 1963, 11 pp.
43. D. R. Douslin and A. Osborn, J. Sci. Instr. $\underline{42}$, 369 (1965).
44. A. G. Osborn and D. R. Douslin, J. Chem. Eng. Data $\underline{20}$, 229 (1975).
45. J. P. McCullough and G. Waddington, "Vapor-Flow Calorimetry." Ch. 10 in Experimental Thermodynamics, ed. by J. P. McCullough and D. W. Scott. Butterworths, London, v. 1, 1968, pp. 369-394.
46. S. S. Todd, I. A. Hossenlopp, and D. W. Scott, J. Chem. Thermodynamics $\underline{10}$, No. 7, 641 (1978).
47. G. Waddington, S. S. Todd, and H. M. Huffman, J. Amer. Chem. Soc. $\underline{69}$, 22 (1947).
48. J. A. Beattie, Proc. Amer. Acad. Arts and Sci. $\underline{69}$, 389 (1934).
49. D. R. Douslin, R. H. Harrison, R. T. Moore, and J. P. McCullough, J. Chem. Phys. $\underline{35}$, 1357 (1961).
50. D. R. Douslin, R. T. Moore, J. P. Dawson, and G. Waddington, J. Amer. Chem. Soc. $\underline{80}$, 2031 (1958).
51. B. E. Gammon and D. R. Douslin, Proc. 5th Symp. Thermophysical Properties, sponsored by the Standing Committee on Thermophysical Properties, Heat Transfer Division, ASME, 1970, pp. 107-114.
52. T. E. Burchfield and L. A. Noll, to be published.
53. M. Mansson, Pure & Appl. Chem. $\underline{55}$, 417 (1983).
54. H. Suga, Pure & Appl. Chem. $\underline{55}$, 427 (1983).
55. C. M. Knobler, Pure & Appl. Chem. $\underline{55}$, 455 (1983).
56. "Selected Values of Properties of Hydrocarbons and Related Compounds," API Research Project 44, Thermodynamics Research Center, Texas A&M University, loose-leaf tables extant, 1983.
57. American Petroleum Institute, Publications and Distributions Section, 2101 L Street, NW, Washington, D.C. 20037.
58. D. R. Stull, E. F. Westrum, and G. C. Sinke, "The Chemical Thermodynamics of Organic Compounds," John Wiley & Sons, Inc., New York, 1969.
59. M. Kh. Karapet'yants and M. L. Karapet'yants, "Thermodynamic Constants of Inorganic and Organic Compounds," Ann Arbor-Humphrey Science Publishers, Ann Arbor, London, 1970.
60. J. D. Cox and G. Pilcher, "Thermochemistry of Organic and Organic-Metallic Compounds," Academic Press, 1970.
61. J. B. Pedley and J. Rylance, "Sussex-NPL Computer Analyzed Thermochemical Data: Organic and Organometallic Compounds." University of Sussex, Brighton, U.K., 1977.
62. T. Boublik, V. Fried, and E. Hala, "The Vapor Pressures of Pure Substances," Elsevier Scientific Publishing Co., Amsterdam, Netherlands, 1973.
63. S. Ohe, "Computer Aided Data Book of Vapor Pressure," Data Book Publishing Co., Tokyo, Japan, 1976.
64. Engineering Sciences Data Unit, 251-259 Regent St., London WIR 7AD, U.K.
65. DeChema Chemistry Data Series, Scholium Publ. Inc., Flushing, N.Y.
66. J. H. Dymond and E. B. Smith, "Second Virial Coefficients of Pure Gases and Mixtures," Oxford University Press, New York, N.Y., 1980.

The Role of the Data Design Laboratory

Joseph R. Downey, Jr. ■ Thermal Group/Analytical Laboratory, 1707 Building, Dow Chemical USA Midland, MI 48640

Process Engineers require a variety of thermophysical property data for engineering design calculations. The role of an integrated data design laboratory in supplying the required data is discussed. In order to be effective, the data design laboratory should have major capabilities in the areas of literature awareness, calculational/estimational methods, experimental methods and critical evaluation. With these capabilities the laboratory can work with process engineers to determine what data are available, what data can be estimated or measured, and whether any or all of the data have sufficient accuracy for the problem of interest. The laboratory will also be capable of working with groups other than process engineers which have a need for reliable estimated or measured data in the thermophysical property area.

The data design laboratory occupies a unique position in the chemical and allied industries. This laboratory shall be defined for our purposes as one which supplies data that forms the basis for process design calculations. In the chemical industry these process design calculations are normally made by process engineers using physical property data, miniplant/pilot plant experience, and a variety of calculational techniques including sophisticated process simulation programs such as ASPEN PLUS*. The results of these calculations can be no more accurate than the physical property/phase equilibria data that was their basis, no matter how sophisticated the simulation process may become. Thus, the successful (safe, efficient and reliable) plant design requies an accurate physical property data base coupled with pilot plant experience and reliable process simulation calculations. The trend in the industry in recent years has been to reduce or eliminate the costly pilot plant and/or miniplant stages whenever practical in order to get production plants on stream faster and at lower overall cost. This places an even greater burden on the reliability of the physical property data and process simulation calculations. This presentation shall focus on the role of the data design laboratory in providing reliable data for these purposes. The importance of accurate thermophysical property data for a number of industrial applications has been reviewed by Sengers and Klein ([1]).

Some of the types of data frequently required for process design purposes are listed in Table 1. This is not an all-inclusive list but is intended to represent the most commonly required types of data. The data design laboratory ideally should be equipped to supply these types of data under all required process conditions. In practice, experimental attainment of this goal is impractical because of the very diverse conditions of temperature and pressure required for some specialized processes. The approach generally adopted by most companies is to set up one or more laboratories equipped to supply the types of data listed in Table 1 over conditions of temperature and pressure commonly encountered in that company's processes. These conditions may vary significantly from company to company due to the nature of the products produced. For example, a company specializing in cryogenic materials may have rather different process data requirements than a company which never deals with cryogenic materials.

In order to achieve its maximum effectiveness the data design laboratory should be far more than simply a laboratory to perform experimental measurements. Instead it should encompass all or most of the following capabilities:

1. literature awareness
2. calculational and estimational methods
3. experimental methods
4. critical evaluation

When configured in this manner, the data design laboratory can work with the process engineers to determine (a) what data is already available in the literature that bears on the problem at hand, (b) what additional data can be estimated or calculated, (c) whether any of all of the above data is of sufficient reliability for the current use, and (d) what additional experimental data is needed and how can it be obtained. Each of these capabilities will be discussed separately in the following sections but the real value of the data design laboratory is the interplay of these activities to solve the problem at hand. The combination of these capabilities in one laboratory results in a synergism which would not be present if they were scattered into several laboratories. For any given problem, one of these capabilities may prove to be more important than the others but, in general, this will not be obvious until the problem is treated in detail.

Also of importance is the impact of the data design laboratory on problems other than those related to process engineering data needs. When configured in the manner described above with emphasis on the areas of thermodynamics/physical properties/phase equilibria of traditional interest for process engineering data needs, such a laboratory is also ideally suited to a number of other data needs within the company. Most notable among these are the reactive chemicals area and the estimation or measurement of properties required for product specifications or for a variety of Research and Development or Technical Service and Development applications.

LITERATURE AWARENESS

This area includes a knowledge of typical sources of data or bibliographies leading to data as well as a general feel for new or improved calculational or experimental techniques. One of the most important aspects is awareness of information sources. These sources are the key to the work which has already been done. The primary scientific literature is the main data source and must be searched thoroughly and completely for each specific piece of data of interest. This search can be quite lengthy and costly and must be done by a skilled searcher. In some cases the literature search may consume more time and effort than new measurements would have taken but this is unusual and typically occurs only if the focus of the search was very narrow; e.g. specific gravity of a single material at 25°C. Broader literature searches on multiple properties of a given material or one property for a variety of materials are very rewarding. The results of such searches form a data base which is valuable input to the process of deciding whether to esimate or measure missing or suspect data.

Searching of the scientific literature is appreciably eased by a number of secondary sources devoted to specialized areas. The disadvantages of these secondary sources are that it may be difficult to determine how thoroughly they cover the field of interest, and they will normally lag or be out of date with respect to the primary literature. Nevertheless, they serve a very useful purpose and the thermophysical property area is blessed with a number of them which greatly simplifies searching of the literature. These secondary sources range from bibliographies, to tabulations of all experimental data, to tabulated data with some correlation to commonly-used equations, to tabulations of selected values, to critically evaluated compilations. Each type has its own advantages and disadvantages which shall not be enumerated here. What is highly important is that the researcher understand which type of source he is dealing with and not mistake one for the other. It could be a serious mistake, for example, to think one is dealing with critically evaluted data when the data was actually taken from the primary literature with no evaluation attempted.

These secondary sources span the spectrum from the very broad range sources such as International Critical Tables,

Beilstein's Handbook of Organic Chemistry or Landolt-Bornstein to more specialized sources which cover narrow areas. Typical examples of specialized sources in the vapor pressure and phase equilibria areas are given as references 2 to 11 and 12 to 21, respectively.

CALCULATIONAL AND ESTIMATIONAL METHODS

The use of these methods are frequently necessary as an addition to or a replacement for experimental data. Even when good experimental data are available it is frequently necessary to represent the data in some equational form for ease of interpolation, extrapolation, or as input to various process simulation programs. Frequently this is accomplished by fitting the data to an equation form using a least-squares procedure to determine the values for the constants. Other methods include calculation of one property from another using known thermodynamic relationships, e.g. enthalpy from heat capacity or heat of vaporization from the slope of the vapor pressure curve. Another important area is the treatment of experimental binary VLE data to obtain the constants for one of the expressions representing excess Gibbs energy, e.g. UNIQUAC, NRTL, Wilson, etc.

Methods for estimation of physical property data have been available for a long time and are continually being improved and expanded. An excellent summary and comparison of many of these methods is given by Reid, Prausnitz and Sherwood (22). Estimation methods are important in cases where it is difficult or impossible to experimentally measure the data of interest and also in cases where a data value is needed quickly for a preliminary calculation with the intent of determining an experimental value to be used at a later date for a more refined calculation. Examples of types of properties which are frequently estimated as a result of difficulty of making accurate measurements are critical parameters and ideal gas heat capacities.

Estimation methods are of various types including group contribution, corresponding states, and homologous series methods. One of the exciting advances in recent years has been the advent of group contribution methods such as UNIFAC (23) and ASOG (24) for the estimation of liquid phase activity coefficients. Although they must be used with care, these methods can achieve high accuracy and they make it possible to predict the vapor-liquid or liquid-liquid equilibria behavior of multicomponent systems from a knowledge of vapor pressure and molecular structure of the pure components. The UNIFAC method has also been applied to calculation of mixture flash points (25) and evaporation rates (26).

EXPERIMENTAL METHODS

The capability of making experimental measurements on systems of interest for process design is of critical importance for the data design laboratory. There is frequently no substitute for direct experimental measurements on the system in question. Of course, it is desirable to have methods readily available with the following attributes: widely applicable, rapid to use, accurate, wide capability in temperature and pressure. As mentioned in the introduction, it is wise to consider setting up apparatus capable of handling many different types of systems. However, there will always be the occasional system which will be inaccessible because of temperature or pressure limitations or corrosivity or toxicity of the chemicals involved. For these cases, one can build specialized apparatus or go to external laboratories which may have the required experimental capability available.

Some of the methods available in our laboratory are outlined in Table 2. The method of choice for vapor pressure is usually the twin ebulliometer in the pressure range 10 to 800 mmHg due to its high accuracy. Other methods available in our laboratory include: vapor-liquid equilibria by an isobaric twin ebulliometric method using an Ellis still or by a semi-automated isothermal static method; cooling curve analysis for sample purity, freezing point, and phase diagram determinations; precision rotating bomb oxygen combustion calorimetry for heat of combustion/formation determinations; high temperature-high pressure flammability measurements in a 36ℓ steel vessel; reaction calorimetry in a number of devices including one capable of operating at the high temperatures and pressures frequently encountered in process conditions. The key to the accuracy of many of these methods is the precise measurement of temperature using precision platinum resistance thermometry (27) or quartz thermometers.

CRITICAL EVALUATION

Critical evaluation is the process of determining the accuracy and reliability of the data and selecting the "best" values. This is essential whether the data was taken from the literature, was determined experimentally for the application of interest, or was estimated. This step is necessary to select the most reliable data from a data set which may contain values of widely varying reliability. Critical evaluation is also necessary even if only one data set is available since its reliability must also be evaluated. Details of the evaluation process would not be appropriate here but major items to be considered include sample purity, details for calibration of measurement apparatus, general accuracy of method used vs. accuracy of other methods, internal consistency of data from each study, agreement with data by other investigators using different methods, and reputation of the investigator for making measurements of high accuracy.

Several examples of critical evaluation activities based on the activities of our group will be briefly discussed. The first of these is the JANAF Thermochemical Tables project (28) which is an ongoing project and has been done by our group under government contract since its inception in 1959. This project involves a critical evaluation of literature thermodynamic data along with the requisite calculations to produce a self-consistent set of thermochemical tables. Originally intended to serve as a data base for military rocket performance calculations, these tables now serve many other applications in process feasibility and analysis in the chemical industry and in energy technologies. While the JANAF tables are primarily limited to inorganic species, our group has constructed a large number of organic thermochemical tables as required for internal company needs. Many of these organic tables were published by Stull, Westrum and Sinke (29).

The second example I would like to mention is more directly tied to process engineering needs. This involves the establishment of critically evaluated pure compound physical property data bases. Several of these are available commercially for purchase or lease including those from Engineering Science Data Unit (ESDU), Thermodynamics Research Laboratory (Washington University), Thermodynamics Research Center (Texas A&M University), and the one under development by AIChE as part of the Design Institute for Physical Property Data (DIPPR). In addition, each company may have a proprietary data base which is based on a combination of literature and proprietary data.

At Dow, a centralized proprietary computerized Physical Property Data Bank (PPDB) has existed for a number of years and now contains data for more than 900 pure compounds. The selection, correlation and entry of data is restricted to two groups of personnel with experience in those operations, but access to retrieve the data is open to all employees. Retrieval is through a number of computer programs designed to serve different purposes. One is a user-friendly program for retrieving tabular data in specified units over desired temperature or pressure ranges, see example in Table 3. The quality codes and temperature ranges are designed to provide the user some indication for the reliability of the data. The other programs which access the data base include programs for distillation column design, the Dow proprietary process simulation program (DOWSIM) and the ASPEN PLUS simulation program. All data are stored in PPDB as coefficients for temperature-dependent equations with emphasis on reduced property equation forms. Coefficients for an equation of state are stored for each compound so that real gas properties may be calculated at any desired temperature and pressure.

DISCUSSION

The data design laboratory can serve a unique role in the chemical industry in terms of its ability to supply high quality thermophysical data function. Of the four capabilities discussed for the data design laboratory, some may be found in process engineering groups although extensive experimental capabilities are rarely found as part of that function. A mode of operation where process engineers serve the literature awareness, calculational/estimational and critical evaluation functions, but rely on other laboratories within the company for required experimental data, is certainly acceptable and may occur rather commonly in the chemical industry.

However, in the author's opinion, an even better mode is to have a separate design data laboratory which incorporate the four capabilities discussed. This results in an integrated laboratory which can serve the needs of the process engineers as well as other data needs in the thermophysical property area.

REFERENCES

1. J. V. Sengers and M. Klein, "The Technological Importance of Accurate Thermophysical Property Information", NBS Special Publication 590, October, 1980.

2. T. E. Jordan, "Vapor Pressure of Organic Compounds", Interscience, New York, 1954.

3. D. R. Stull, Ind. Eng. Chem., $\underline{39}$, 517 (1947); ibid $\underline{39}$, 540 (1947).

4. S. Ohe, "Computer Aided Data Book of Vapor Pressure", Data Book Publishing Co., Tokyo, 1976.

5. T. Boublik, V. Fried and E. Hala, "The Vapor Pressure of Pure Substances", Elsevier, Amsterdam, 1973.

6. J. A. Riddick and W. B. Bunger, "Organic Solvents: Physical Properties and Methods of Purification", Wiley-Interscience, New York, 1970.

7. J. Timmermans, "Physico-chemical Constants of Pure Organic Compounds", Elsevier, New York, Vol. 1, 1950 and Vol. 2, 1965.

8. R. R. Dreisbach, "Physical Properties of Chemical Compounds", ACS, Washington, DC, Vol. 1, 1955; Vol. 2, 1959; Vol. 3, 1961. These are Advances in Chemistry Series #15, 23 and 29, respectively.

9. Engineering Sciences Data Unit, "Physical Data, Chemical Engineering Series", in 8 volumes, current.

10. K. R. Hall et al., Thermodynamics Research Center Hydrocarbon Project, "Selected Values of Properties of Hydrocarbons and Related Compounds", Texas A&M University, current.

11. K. R. Hall et al., Thermodynamics Research Center Data Project, "Selected Values of Properties of Chemical Compounds", Texas A&M University, current.

12. J. Gmehling, U. Onken et al., "Vapor-Liquid Equilibrium Data Collection", DECHEMA Chemistry Data Series Volume I (in 7 parts).

13. J. M. Sorensen and W. Arlt, "Liquid-Liquid Equilibrium Data Collection", DECHEMA Chemistry Data Series Volume V (in 3 parts).

14. H. Knapp, R. Doring, L. Oellrich, U. Plocker and J. M. Prausnitz, "Vapor-Liquid Equilibria for Mixtures of Low Boiling Substances", DECHEMA Chemistry Data Series Volume VI, 1982.

15. E. Hala, I. Wichterle, J. Polak, and T. Boublik, "Vapor-Liquid Equilibrium Data at Normal Pressures", Pergamon, Oxford, 1968.

16. J. J. Christensen, R. W. Hanks and R. M. Izatt, "Handbook of Heats of Mixing", Wiley-Interscience, New York, 1982.

17. I. Wichterle, J. Linek and E. Hala, "Vapor-Liquid Equilibrium Data Bibliography", Elsevier, Amsterdam, 1973; three supplements are available from the same publisher.

18. A Marzynski et al., "Verified Vapor-Liquid Equilibrium Data", PWN-Polish Scientific Publishers, Warsaw, in 6 volumes.

19. J. Wisniak and A. Tamir, "Liquid-Liquid Equilibrium and Extraction", Vol. 7A, 7B, Elsevier, Amsterdam, 1980; a bibliography.

20. J. Wisniak and A. Tamir, "Mixing and Excess Thermodynamic Properties", Elsevier, Amsterdam, 1978; a bibliography.

21. H. Stephen and T. Stephen, "Solubilities of Inorganic and Organic Compounds", Vol. 1 and 2, Pergamon, Oxford, 1954. Volume 3 of above series was authored by H. L. Silcock and published in 1979.

22. R. C. Reid, J. M. Prausnitz and T. K. Sherwood, "The Properties of Gases and Liquids", 3rd ed., McGraw-Hill, New York, 1977.

23. A. Fredenslund, J. Gmehling and P. Rasmussen, "Vapor-Liquid Equilibria Using UNIFAC", Elsevier, Amsterdam, 1977.

24. K. Kojima and K. Tochigi, "Prediction of Vapor-Liquid Equilibria by the ASOG Method", Elsevier, Amsterdam, 1979.

25. J. Gmehling and P. Rasmussen, Ind. Eng. Chem. Fundam. 21, 186 (1982).

26. A. L. Rocklin and D. C. Bonner, J. Coatings Technol. 52, 27 (1980).

27. D. R. Stull, Ind. Eng. Chem. 18, 234 (1946).

28. D. R. Stull and H. Prophet, "JANAF Thermochemical Tables", 2nd ed., NSRDS-NBS 37, U. S. Government Printing Office 1971; supplements J. Phys. Chem. Ref. Data 3, 311 (1974); ibid 4, 1 (1975); ibid 7, 793 (1978); ibid 11, 695 (1982).

29. D. R. Stull, E. F. Westrum and G. C. Sinke, "The Chemical Thermodynamics of Organic Compounds", Wiley, New York, 1969.

TABLE 1

Frequently Needed Process Design Data

Vapor Pressure
Heat of Vaporization
Specific Heat/Enthalpy
Density
Thermal Conductivity
Viscosity
Surface Tension
Phase Equilibria
    Vapor-Liquid Equilibrium
    Solubilities
Heats and Kinetics of Reaction, Polymerization, Solution, Dilution, etc.
Flammability Hazard Data
Reaction Hazard Data

TABLE 2

Experimental Methods in Use at Dow Chemical Thermal Group

Type of Measurement	Method	T Range, °C	Sample Size	Accuracy	Time Required, Day
Vapor Pressure	Twin Ebulliometer	amb to 400	15 ml	<0.1%	1/2
Vapor Pressure	Static Manometer	amb to 250	2 ml	0.2%	2
Vapor Pressure	DSC	-180 to 750	2 mg	1%	1/2
Sp. Heat/Enthalpy	DSC	-180 to 750	2 mg	2%	1/2
Thermal Cond.	Transient Hot Wire	amb to 250	20 ml	1%	1/2

amb = ambient temperature

TABLE 3

DATA FROM THE DOW PHYSICAL PROPERTY DATA BANK - 08-02-83

471   O-DICHLOROBENZENE   REV. 3/11/81
         C6H4CL2

MOLECULAR WT	147.0036		TC 417.200 DEG C	Q=1
BOILING PT	180.395 DEG C		PC 30232.8 MMHG	Q=1
FREEZING PT	-16.970 DEG C	Q=5	VC 2.43531 CC/GM	Q=1
MOL DIF VOL	128.340	Q=1	DC .410625 GM/CC	
ACENTRIC FCT	.305392		ZC .251390	

TEMP.	VAP PRES	DENSITY, GM/CC		HEAT CAPACITIES, CAL/GM-DEG C			
DEG C	MMHG	SAT LIQ	SAT VAP	CS LIQ	CP LIQ	SAT VAP	IDEAL GAS
120	132.45	1.192	.8044E-03	.3117	.3117	.2311	.2293
140	252.19	1.168	.1469E-02	.3194	.3193	.2406	.2377
160	448.53	1.143	.2518E-02	.3272	.3270	.2502	.2457
180	752.66	1.117	.4095E-02	.3351	.3349	.2600	.2534
200	1201.6	1.090	.6371E-02	.3433	.3430	.2701	.2607
220	1837.9	1.062	.9553E-02	.3517	.3513	.2806	.2677

		----------- E N T H A L P Y , CAL/GM -----------			
T, C	VP, MMHG	SAT LIQ	VAPORIZ	SAT VAP	IDEAL GAS
120	132.45	-18.5146	70.5681	52.0536	52.3072
140	252.19	-12.2038	68.7564	56.5526	56.9784
160	448.53	-5.73860	66.8802	61.1416	61.8138
180	752.66	.885498	64.9123	65.7978	66.8059
200	1201.6	7.59178	62.9070	70.4987	71.9477
220	1837.9	14.4429	60.7789	75.2218	77.2322

TEMP	VISCOSITY, CP		THERM COND, CAL/SEC-CM-C		SURF TENS
DEG C	LIQUID	LOW P VAPOR	LIQUID	LOW P VAPOR	DYNE/CM
120	.5078	.9804E-02	.2510E-03	.2179E-04	26.37
140	.4458	.1029E-01	.2430E-03	.2379E-04	24.35
160	.3972	.1078E-01	.2350E-03	.2582E-04	22.35
180	.3583	.1127E-01	.2271E-03	.2788E-04	20.38
200	.3267	.1175E-01	.2193E-03	.2997E-04	18.44
220	.3006	.1224E-01	.2116E-03	.3207E-04	16.51

		DATA RANGE				DATA RANGE	
	Q	DEG C			Q	DEG C	
VAPOR PRESSURE	5	0	182	VAPOR VISCOSITY	1	0	360
IDEAL GAS CP	5	25	1227	LIQUID VISCOSITY	5	-18	175
LIQUID CP	5	-17	140	VAPOR THERM COND	1	0	360
HEAT OF VAPORIZ	4	0	182	LIQUID THERM COND	3	-10	80
LIQUID DENSITY	5	0	150	SURFACE TENSION	3	20	40

DATA QUALITY CODES (Q): 0=UNKNOWN, 1=ESTIMATED, 2=ONE SOURCE-WEAK,
3=ONE SOURCE-OK, 4=ONE SOURCE-GOOD, 5=CHOSEN FROM MULTIPLE SOURCES

PANEL DISCUSSION: ARE WE EFFECTIVELY COMMUNICATING INFORMATION TO CHEMICAL ENGINERS?

T. Selover, Chairman

MAX KLEIN (Gas Research Institute, Chicago, IL) - When an attempt is made to omit the pilot plant stage and design a plant from data base properties it is the accuracy of the data used that is important and not simply their consistency. The danger inherent in this process of design without a pilot plant stage is that when all the different kinds of data are put together, each with its own errors, an operating plant may not result. This problem becomes exacerbated in the process of predicting values for data needed but which are not available in the data base. All too often, under pressure of the design deadlines, predictions are based on inadequate models which are often used beyond the range of even their approximate validity. Yet, as Neil Olien pointed out in his talk, the determination of accurate data by measurement alone in response to design need is unrealistic. Because there are so many systems of interest and so many operating conditions for each system, the cost of obtaining data through measurement alone becomes far too costly. Even were this not so, the time involved in making such measurements would preclude this approach. The designer must, therefore, make maximum use of the existing literature and of such predictive models as are available. Yet, many within the chemical engineering design community are unaware of the large number of data sources that are available to them. It was this observation by the organizers that led to the design of this symposium.

We wish now to open this session to a discussion of this problem of how to promote the broader and more efficient use of existing literature data among chemical engineers. I would like my co-chairman, Ted Selover, to introduce the discussion and act as chairman.

TED SELOVER (SOHIO, Cleveland, OH) - The title of this panel discussion, as it appeared in the meeting program, is somewhat vague and says, "Are we effectively communicating information to chemical engineers?" I wish first to develop a framework for the kind of discussion we hope to develop in this round table. We wish to determine if those who produce data and their colleagues who accumulate those data into data bases feel that potential users of their work are actually finding the data or are even aware of their existence. Secondly, do these users know what they can do with the data once they get them. In other words, is there a traceable link between the person who produces the data and the one who needs and uses the data. In thinking of the answers to this question or in framing your discussions on it, I wish you would try to determine how it is possible to improve on the use of this kind of information through several of the different means available to us. Some of these means have already been discussed, but I will mention them as examples of what might be useful topics for discussion.

Consider, for example, what is referred to as the gateway network. Operationally, this can involve either computerized on-line output or hard copy. A news bulletin format is needed in this kind of network, like that of OSRD, CODATA or the CRC tables. Such a bulletin would list what is available, who has data and where the user can go for help in the use of the data. A middleman, who can be called the information gate keeper, is very useful in such a system. This is a role that I happen to fill at SOHIO. This middleman knows what is going on in data generation and accumulation and can transmit the data developed from the data generator to the users who need data.

We could also discuss one of my pet subjects. That is, the need for the renewal of teaching the use of technical literature at the college level. It seems that this subject has been almost totally eliminated at most science and engineering schools. Yet, this could be the key needed for getting people aware of what is going on in the world of data.

With these remarks as a framework for a good discussion, let us now open the floor to people who want to elaborate on their ideas.

NEIL OLIEN (National Bureau of Standards, Boulder, CO) - What I want to say may be taken as the antithesis of what I was promoting in my talk. Some of the things that were said yesterday and then again this morning are of considerable concern to me. I refer to the current enamorization with modeling. This was referred to in the talk about ESDU this morning. It was said that engineers nowadays need only sit in front of a terminal and call up data. I do not really believe this to be true. I am deeply concerned with the fact that those at the management level in companies and research laboratories have become enamored with modeling. Yet, as I strongly emphasized in my talk, experiment must soak up the bulk of the funding available for data. I am very much concerned that, in the various companies, decisions will be made to put most of the money available into modeling because of the impression which management has been given that more bangs for the buck can be so obtained. This approach cannot help but result in a further decline in the availability of experimental data. We will then find ourselves in a never-never land modeling systems based on the use of methods which have no basis in reality since they are ultimately based on no or, at best, inaccurately measured data. I am very concerned about this. We must somehow get across to the engineer - user and to his management that it is only possible to model systems when there is a large set of very accurate data behind the model and when those data have been obtained by people who know what they are doing. This is part of my wide-ranging concern for the current decline of experimental capability in this country. Our country went through such a decline about 15 years ago. This decline essentially ended about five years ago. I now hear management saying again that measurements are no longer needed but that all data that are needed can be obtained from a data center like PPDS, i.e. the Physical Properties Data System in Glasgow, Scotland which was referred to in a talk this morning. People at that center and at similar ones do not make measurements! I am concerned when I hear upper management say that measurements are not needed and that all the data that are needed can be obtained from someone who does not make measurements. I am not sure if my remarks are appropriate for this discussion, but that is a problem that is currently very much on my mind.

TED SELOVER - That is indeed a good comment and a very important one. I would be interested in hearing from anyone in the audience who has made a parametric study of the sensitivity of a plant design to errors in the data used. Take, for instance, the use of critical point data in design. Such data are the building blocks for all corresponding state correlations. In the DIPPR program, it was discovered that critical point properties are not known very accurately for many substances. Furthermore, how to measure them is not always clear. SOHIO built a plant to produce 400 million pounds of acrylonitrite a year despite the fact that the critical properties of that material are not known. This could have led to a very bad situation. The plant was built, and fortunately it works. I suppose we were lucky. I will also have to make one comment in answer to Neil. Of course, the Physical Properties Data Service does not make measurements, but the National Engineering Laboratory, where they are located, is well known for the measurements that it makes. This is especially true for some of the properties that are critical in the engineering business. It is those data, measured at NEL, that are often used in the information that PPDS publishes. I have no special reason to be their advocate but felt that I had to make that point.

MAX KLEIN - There is a distinction that needs to be made between measurements made for developing fundamental models and measurements made because somebody needs a set of data in a particular range for design purposes. It is my feeling that the situation corresponding to the latter leads to an unending and very inefficient method of data production. I cannot overemphasize the fact that measurements are essential in the development of models. Unless a model is based on properties of the most elementary particles, it must contain various kinds of molecular parameters which need to be determined from experimental data in order to develop what the truth is. Because it is so expensive and takes so much time, experimental data must be taken for purposes beyond immediate design needs. It must supply us with knowledge, usually through its use in the development of models, which will lead to a future ability to predict properties instead of having to take yet another set of measurements for the next design need. Only in this way can a measurement program be truly cost effective in the large sense.

TED SELOVER - There has been a great deal of discussion of gateway networks. These are networks which deal with each other and with data bases. Bill Thompson and I had some interesting discussions on this between sessions. Bill, would you care to elaborate concerning what you think should be done here.

BILL THOMPSON (Royal Military College, Kingston, Ontario, Canada) - This noon we discussed the fact that many people are not aware of the various data banks that are available. One step that could be taken to make them more aware of such data banks would be to develop and advertise a hotline number which listed all of them. In fact, the discussion we had at lunch related to my comment on my experience with this meeting. On one of the pages in the AIChE meeting program there is, in fact, listed a hot line. I used that number to facilitate my travel arrangements, my registration at this meeting, etc. It made everything much easier with everything taking only five minutes. Such a number could be established for data bases. It could be advertised as a telephone number available for anyone to call from his computer terminal in order to get a list of the various resources presented on his computer screen. This would certainly be a step in the right direction toward a more widespread use of all data bases. This could eventually lead to the development of a complete gateway system such as Jack Westbrook and Jim Graham described this morning. I talked to Jack about this earlier this afternoon and learned that something along this line is already under way involving Viktor Hampel at Livermore. Perhaps he would like to elaborate on where that particular activity now stands.

JACK WESTBROOK (General Electric, Schenectady, NY) - There are two different concepts here. One involves having an on-line network such as Jim Graham and I each described where it is possible to get in touch with the actual data base by a single telephone call. The other is appropriate for those data bases, whether machine readable or not, which, whether for proprietary reasons or for software or hardware compatibility problems, cannot be reached directly. It would be helpful for interested users to know about their existence. It would be useful to be able to call on the telephone, either by voice or through a terminal, and be made aware, for example, that Dow has some particular data base on certain organics and NBS has one on something else, etc. Both of these approaches need developing.

At a workshop in Tennessee in November 1982, I reported that I had identified some 58 data bases on material properties in the MPC study which came into being since the Hilsenrath NBS publication of November 1980. Since then, the three of us, that is Hampel, Hilsenrath and I, along with some of Hampel's coworkers from Lawrence Livermore, have put these together and have even added some we became aware of since. There are now more than 120 machine readable materials properties data bases with which we are familiar. Viktor gave a paper on this at an ASME meeting in Chicago earlier this month. That paper is available in hardcopy form. Those of you who want a copy can reach Viktor at Lawrence Livermore. We are now in the process of putting the file on-line. It will then be possible to dial Viktor's computer and see this entire set on your terminal. Given sufficient patience, it will be possible to scroll through 120 data bases and see everything. If, on the other hand, there is only an interest in, say, the thermodynamic properties of organics, it will be possible to zero in on the 3 or 4 or 5 that relate to that particular area.

We would also like to to be able to take the various hardcopy guides, such as the one that I mentioned this morning that appeared in the

Annual Review of Materials Science with many hundreds of references and put them on line. This is a much bigger job than what we have done and will require some special funding and thus far I don't see any funding for this. Another point that Bill Thompson and I discussed a bit earlier relates to the fact that getting these bases available to a wide group of people and accessible to a single phone number is not just an assist to the user, although it certainly is that. Perhaps the greatest ultimate benefit which will come out of this will be the development of an interactive system. For instance, a particular user might look at the file for this system and see that his data base was omitted. This is, in fact, just the kind of reaction we would like to stimulate. After all, it is very useful to hear that someone has or knows about another data base some place whether it is on-line, machine readable or just in print. This is often very hard to determine through any normal literature searching procedure. It is an added benefit that will be available when the two steps indicated are taken to develop directories.

TED SELOVER - Jim Graham can you comment on the feasibility of the extension of your networking concept to thermophysical properties, fluid properties, etc.? In other words, how open are you to including this type of data in your data base?

JIM GRAHAM (Deere & Co., Moline, IL) - We are quite open to such additions. In fact, one of the best discussions I have had with another data base supplier was one I had with CINDAS. Dr. Ho, who is now the director of CINDAS, is quite interested in becoming a part of the network that the MPC is developing. I see no real technical reason why this cannot be done, although there may be a different format required for other properties. I think the concept can be extended to a large number of other types of property information.

TED SELOVER - Do you see that this will follow a sequence in which the materials property development is completed first and these other areas added on, or do you think it could all develop in parallel?

JIM GRAHAM - I think it might develop in parallel. We have another meeting in Boston this Fall to talk about material properties at ASME. One of the speakers at the meeting is to be Dr. Ho, and he plans to make a strong pitch for CINDAS to join the MPC effort. We will probably develop the formatting and some of the standards for the data bases in parallel. NACE, the National Association of Corrosion Engineers, also has an interest in coming in. They attended our last committee meeting and expressed an interest in participating with us. NACE has established a committee that will begin to generate some of the data to be put into the data base. I think it is possible to open this up to a large variety of properties.

MAX KLEIN - I want to ask Jack Westbrook if DIALOG or Engineering Index might be interested in setting up a user file made up of the titles of these data centers along with an abstract from whatever publications there are which describe them. Those computer search organizations might consider there to be enough money in this problem to set up a file that could be searched with key words in the manner that Chemical Abstracts or Engineering Index are currently searched. That means that those companies would do the conversion to the machine. Have you thought about that or tried it?

JACK WESTBROOK - I have been in touch with all these groups perhaps not on that particular point but on getting them to put the numerical data themselves on line. I have found them to be reluctant to do that. I know you are referring to something else, but my contact with with them was on the data base. They were not interested for two or three reasons: 1) they did not feel they had the expertise on how to decide on how to express the properties, on what to do about limits of utility, and on how to decide if numbers are good; 2) they were doubtful if the market was large enough to sustain the expenses. The numerical bases with which they are currently involved have been confined to the business and socio-economic areas. I did not ask them about the specific point you mention, that of developing a referral center either for data bases or printed compilations. That is a worthwhile suggestion and they might have a different attitude toward it.

MAX KLEIN - A file of data bases would be analogous to what they have now. Currently, the user enters their system with key words and the system searches a title and abstract, or so I believe. The user is then given information which identifies the original document to which he can go if he so desires.

In this case, this would be the procedure for going to the original data base.

TED SELOVER - Jack did you say that CRC had something on BRS on-line listing these different data bases.

JACK WESTBROOK - What they have is different again. CRC Press has what they call a super-index. CRC Press publishes hundreds of handbooks and monographs of all sorts. Recently these have been mainly in the life sciences but have been in many other areas as well. What CRC did was to take the indices from all of these and "mash" them together into one index, not only their own but those of various other publishers as well. Let me indicate some of the problems they ran into. For example, they found that if they asked for plasma, the super-index would refer them to several papers or books. In one case it might be a reference to blood plasma and in another case it might be to an electron plasma. In each case the word came from a different book, and in this case, two different disciplines. Another problem can be illustrated by a made-up example. Suppose the thermal coefficient of expansion of neodymium is desired. The request is entered and a page number appears. Suppose further that when the data base is called on line, it is found not to be there. On checking to see if the correct table was called up, it was found to be it is the right page. What is the trouble? There is a column marked neodymium on that page and a row marked thermal expansion, but the entry is blank. Thus, neodymium was indeed on that page, but the index refers to an empty set. Thus far CRC has not been able to cope with this problem. This is a good idea but obviously still has bugs in it.

DR. CHANG - I would like to speak to this group from a user's point of view. I have developed a commercial process modeling program for petrochemical technology. I have used a thermodynamics data base for fugacities, etc. I would like to offer you my experiences. I have used over twenty components with thermodynamic functions and properties. It is very important to recognize that everything depends on the degree of sophistication of the process modeling program that the client requires. When more detail is needed, my experience is that the user and compiler will call for more complex thermodynamic properties. The line of communication is therefore very important. It is the demand for accurate process modeling that govern the needs of these services. We need more thorough thermodynamic manuals as the demand for modelling services grows. I just wanted to point out that this has been my experience. With the advent of on-line information, including satelite communications, it has become possible for us to determine if we should use more detailed process modeling for a commercial operation.

TED SELOVER - I would like to have Bill Kirchhoff comment on the CIS Chemical Information System data base and the possibility that maybe a listing of these different physical property data bases or information sources could be made available on CIS. This is a data base that many chemists use for spectroscopic data and environmental problems.

BILL KIRCHHOFF (National Bureau of Standards, Gaithersburgh, MD) - CIS certainly seems to be a very flexible data base in that new properties and the contents of new data centers are added when the person running the data center feels that he wants to participate. I would prefer to comment on a more general topic at this point, however. It seems that we are at something of a crossroads in data handling at this point. It still costs much more to get a datum out of a computerized system than it does to buy a hardcopy book that has all the data on the property of interest. I am always fascinated by the fact that people are willing to spend $50-$100 in computer time or hookup time to get the information out of the computer, but for some reason, find it difficult to justify the expense of that same $100 to buy the entire hardcopy version from which the computerized data base was, in fact, generated. It is certainly clear that at sometime in the future all data will come from computerized data bases. These will contain both measurements and calculations of properties from models. They will also be interactive in a much broader sense than I have heard discussed here today. There is an opposite communication that has not been addressed at this symposium and that is how the chemical engineer gets to the people who are in charge of data centers to tell them what his or her needs are. There are many such users. In the past, when NBS sold its handbooks Circular 500, or Technical Note 270, they did not know who used them and for what kinds of data and whether the user was happy with what he found. They did not even know if the user agreed with the numbers that were in the publication. With these on-line systems, when a user wants to call in and get a particular property over a particular range and finds it not to be there, he will be able to flag the people who are operating the center in such a way as to tell them that there are data needed where they do not exist. It is a very good way to find out what the data needs of the broad engineering community are. This could then be channeled to the funders of research who might then fund measurements or the calculation of additional values.

The major advantage of computerized data bases is that all of the information of interest, subject to proprietary considerations, might be available from the computer terminal in the future. To be really successful, such data bases would have to be very large. They have to be operated by chemical engineers, theoretical chemists and those experienced in making measurements. On the other hand, I am afraid there will always be a tendency for such large data bases to be run by computer people who like putting together large data bases with very complicated interactive programs. I do not believe that we currently have a single system that has enough memory, enough of a diversity of people involved with it, and enough of a reputation to reach the kind of efficiency of scale that seems to be lacking in all of the systems that we have heard described today.

The system that will be the most successful in the future is the one that will be sufficiently flexible to take advantage of user input so as to be able to adapt to new needs. The Chemical Information System is certainly one that has that kind of potential. There are a large number of government agencies with big computers involved. It seems to be growing and expanding as new data centers become involved whether with environmental or health issues. Those particular issues were most important in the beginning of the Chemical Information System. They no longer seems to be quite as important. I know that thermodynamic properties are now going into the system. Cystallographic data, which as far as I can see has no relationship to toxicology or to environmental health, are also going into it. The CIS may be the system that is well worth watching. Anyone with a computerized data system might want to get involved in the CIS in some way. As I understand its operation, it really is the coordination of independently developed and independently paid for computerized data systems. Suppose I were running a data center in microwave spectroscopy, as I once did, and put my data center into that system. Then, if someone accessed my data center through CIS, there would be a mechanism for paying a user fee back to me after the central system took off its share from the user charge for the management of the system. This may be the system that will grow although it may be developed in another country. With the advent of satellite communications we are not going to be restricted to land lines of communication for very long in the future. I am not sure exactly where the future will take us but I do think CIS is only one possibility.

<u>TED SELOVER</u> Good answer.

<u>Bill KIRCHHOFF</u> - If no one wants to respond to that, I will make a slight change of direction. There is one other point that I would like to make because I am very much concerned about it. This has to do with the reporting of uncertainty limits and the teaching of the proper error analysis methods. This would result in the proper incorporation of uncertainties into data bases that consist of measured and reported values as well, as in the incorporation of uncertainties into computer models. Of course, I really cannot expect anyone to take an admonition like this very seriously because I find myself becoming very cynical as I get older. I think everyone pays lip service to the reporting of uncertainties, but none really seems to do anything about it until the situation gets to be terribly critical. I would urge people who have the responsibility for educating youth to teach them to estimate the errors in their own data properly and to seek constantly for the sources of error in reported data. I would strongly urge people who have computerized data bases or who publish books of data not to record a single number from any source whatsoever without an estimate of its uncertainty. I think that as I turn into a curmudgeon, my hair falls out, and I become really grouchy, I will start making a personal career out of becoming really snotty about that sort of thing. I am ashamed to say that it was a congressman who was giving a keynote speech at a conference that I organized who made the very simple statement that a number without an estimate of its uncertainty is not worthy of being recorded. That is a truism that we simply forget all too often. This is an admonition to everyone here, including myself.

JOHN DEWEY (Reynolds Metals Co., Richmond, VA) - I have been concerned about this uncertainty problem and with how an estimate of uncertainty is obtained. I can cite a case from a recent paper of mine containing some data and a correlation for the solubility. By doing just a straight correlation without fixing anything, I can get somewhere close to the literature values of the enthalpy change of that reaction at standard conditions. I can vary that enthalpy and put it in as input. I then obtain T values and some other coefficients that do not vary. The coefficients themselves can vary quite dramatically. At this point all I could say was that value of the enthalpy has been determined by respectable researchers and I obtained results within their claimed uncertainty. I fix the enthalpy and get T values from the other coefficients. Now I do not know what else to do. In the publication, I do believe the T value or the standard error should be marked on the data, and I do that. Other than that I do not know what to do.

BILL KIRCHHOFF - I did not say that determining the uncertainty was an easy thing to do. I certainly do not believe that there is a formula that can simply be followed by rote. As a starting point, I would just urge that the author at least state the procedure he used in estimating his uncertainty and make his best stab at doing it.

JOHN DEWEY - I must agree with that. One of my pet peeves over the years has been that people publish regression analyses and never put the standard deviation of the coefficients on it. As far as I'm concerned that is worthless as a publication.

BILL KIRCHHOFF - I think the journal editors and referees probably all have a responsibility to try to get the scientific community to stick to such higher standards.

TED SELOVER - Do we have other pressing issues or comments that people want to make on any of these subjects?

JACK WESTBROOK - In introducing the round-table Ted, you raised the issue of computerized versus hard copy. This has been followed by discussion containing the implication that hard copy would eventually disappear and would become something that would be found only in a museum. In other words that all information would be obtained from a computer. I see it as a bit different from that. I think that priorities probably will change. Currently, it is mainly the well-known, well-loved and well-used hard copy data bases that are becoming computerized and are thereby being given new capabilities through that process. On the other hand, we heard of several instances today in which the hard copy will soon be derived from the computer file. This will be very much handier for the user. Thus, instead of having to have a personal library of several hundred key reference volumes, with only 5 or 10 percent of the contents of each being useful, it will be possible to make handbooks customized for the user. This will be done simply by taking data dumps from various online bases including only relevant portions and putting them together. This does not come free, however. It will probably cost more than the cost of all the current handbooks. It will, however, result in a customization opportunity that was never available before. This can also be customized in another way. Thus if, for example, vapor pressure values are preferred at round number temperatures, they will be so available on demand. On the other hand, if the temperature for specific values of the vapor pressure are desired, that will be available on demand. This cannot be done with a handbook. Those yield the format the editor and compilers chose and it is necessary to live with those formats. Thus, the possibility of deriving hard copy from the computerized file will lead to a whole new family of hard copy publications that are different than what is available today.

MAX KLEIN - I want to make a comment on the issue of hard copy versus computer copy. I agree with Jack Westbrook that this is the way things will go. One thing that this will cause you to lose, though, is the serendipity you now have. Thus, when you now look for a number, you see a set of numbers and you see a trend. Often you see something that you hadn't expected to see which is useful. This will not happen when you go to a data base for a simple number. It may be possible to display some of the data graphically in order to preserve some of this, but there will always be something that will be lost. The browsing aspect of looking at other numbers and other information in the more general hard copies will disappear.

TED SELOVER - At this point, I wish to comment on DIPPR. We have not heard from DIPPR, and since I have a responsibility in their public relations area, I should give a spiel here. One of the projects that DIPPR is supporting is one at Penn State on data compilation. The original product from this project is a set of data sheets which contains all of the constants of temperature invariant properties, along with the coefficients to generate all the temperature dependent properties. There are approximately 38 such properties. The users and project supporters have told us that they do not want data supplement sheets. It seems that the process of throwing out the old ones and putting in the new ones in the notebook is a terrible logistics problem. What they want is a computer output each time. The DIPPR project does have a way to generate computer output from the source file and the supporters of this project are going to do exactly what Jack Westbrook says. That is, they will make their own tables from the equations, using only the constants that are stored. The original hard copy form will become archival. It then becomes possible to develop your own data base from the DIPPR data tailor-made to your own needs. When this is available to the public we hope you will all buy it and use it. Is there a comment on this or anything else?

BRUCE GAMMON (National Institute for Petroleum and Energy -NIPER, Bartlesville, OK) - I think that someone may have asked this question, but I want to ask it again. Has anyone estimated the volume of material and the kind of mass storage media needed for the kind of data bases that you envision? What is it going to cost to maintain that kind of mass storage capability?

JACK WESTBROOK - The answer briefly, is yes. We have estimated that, and of course, the answer depends on how the data bases that you are talking about are defined. It is fortunate that the available memory capacity that can be put into a reasonable amount of space has been growing ever larger, while the cost per unit of information stored has been dropping precipitously. Thus, while this might have been an issue and a serious problem five years ago, it is hardly an issue today. It certainly will not be one in another five years. People just do not have to worry about that very much anymore.

NEVIN GOKCEN (U.S. Bureau of Mines, Albany, OR) - I have three unrelated remarks. One of them relates to what Jack Westbrook just mentioned. My feeling is that we shall have a happy balance between computers and hard copies. There is, in my mind, no question about that. Some users will find hard copies better than computer output. Some may not have access to a computer, and others will not have access to a computer during certain times. There is a three hour time difference between the East and West Coast, and that alone is a problem when someone on one coast needs to use a computer on the other. My second point deals with handbooks. I still use the Chemistry Handbook, and in fact, I still use Linke and Seidell's Solubility Handbook. I think there is a need for handbooks of this type and for a very important reason. There is a need for selected values of thermodynamic or thermophysical property data which are not necessarily comprehensive. In order for the user, especially for the chemical (or other) engineer user, to proceed at a fast rate, he needs selected data that take much less time to generate or compile than do those from the comprehensive approaches such as the ones we have been producing or that Dow Chemical or the Bureau of Standards have been producing. That is my second remark. My third remark is really a comment. I see very little technical interaction between industry or the practical engineer and the research centers that are producing data and data bases.

I hereby volunteer to you that anyone who needs thermodynamic data for inorganic compounds, metals, and elements should simply call Bureau of Mines at 503/967-5866. I also have a corollary to this that I would like to state. I am not sure if it is because of their educational backgrounds or because they have so many subjects as undergraduates or graduates, but engineers exhibit resistance or fear in using thermochemical or thermophysical property data. To help people with this problem we would be happy to make ourselves available after computations are made from the data obtained from us or the Bureau of Standards. We would be very willing to answer a call asking us if any given calculation seems correct or not. We would certainly be ready to help out. Thank you.

TED SELOVER - Now, I would now like to hear from some of the academicans or anyone else who feels strongly about the absence from curricular of courses on chemical information or engineering information or whatever other name is appropriate. Such courses do not seem to be taught any more in universities. If there were someone teaching this material to students, they would move into industry with an awareness of where data are to be found and how to get at data. More people might then become information gatekeepers because they might find that an attractive field to enter. It appears that no one wants to teach such a course. It was certainly taught by somebody before! What happened to the people that used to teach it? I don't understand why academic people say they do not have enough place in their curricula to teach these courses that we are discussing on the use of the literature. Such courses used to be in the curriculum and it is still always said that the most important thing in design is to use good data. Often going to the Handbook of Chemistry and Physics for a property of a hydrocarbon such as the free energy of formation yields three different values for the same compound at one temperature. All three numbers are different because they came from three different tables. Yet no one seems to worry about this problem. Neither courses on where to find good data nor how to distinguish good data from the bad data are taught. As a chemist it bothers me that engineers just do not want to face up to this issue.

ALLEN MATHER (University of Alberta, Edmonton, Alberta, Canada) - Certain kinds of problems are not covered anymore hence many tech- niques have been dropped from the curriculum. There is also much more that has to be put into the curriculum and things that are regarded as peripheral have been dropped. Searching and using the literature is one such item that has been considered to be periph- eral. So have many other subjects been dropped and to our disadvantage. Many writing courses were dropped and this is also unfortunate. Where I see we can perhaps do something about this is on the graduate level. The beginning of many graduate programs should perhaps include an introduction, perhaps not a formal course but an introduction, to using literature and searching for data. At the undergraduate level I don't see that it is possible to do this, except perhaps as part of lab courses. I do not, however, see this as a formal course in the curriculum, in any event.

TED SELOVER - As my last comment on this, I would like to say that I thought this was the most important course I ever took in undergraduate school. I did not get any credit for it, yet I had to take it.

BOB FREEMAN (Oklahoma State University, Stillwater, OK) - At the risk of making our curriculum seem old-fashioned, I will tell you that anyone who has students who would like to go through a major in chemistry or who want to get a graduate degree in chemistry and take such a course should steer them to us at Oklahoma State University. We offer one.

TED SELOVER - Great!

BOB FREEMAN - As long as I am up here I'll make one other comment relating to Bill Kirchhoff's comment on uncertainty statements. I wish simply to give as a demonstration as bad an illustration as I have ever seen of the point he made. There are two commissions from IUPAC jointly working on the revision of Latimer's tables of half cell potentials. I had occasion recently to see some sample sheets from that effort. In their tables, they list half cell potentials to something like 10 microvolt. This is 6 significant figure in many cases. Yet, nowhere do they talk about any uncertainty. One of the coeditors of that publication is a major editor of an ACS journal.

TED SELOVER - Your point is that the data on which the half cell data are based are perhaps good to two significant figures and yet there result six in tables.

BOB FREEMAN - I do not even know how good the data are. It is not possible to tell.

TED SELOVER - They had better take Bill Kirchhoff's course in significant data.

BOB FREEMAN - You can assume that the data are probably as good as two of the figures given but I sure wouldn't bet on it.

TED SELOVER - Any more comments.

JACK WESTBROOK - While we are telling horror stories, I will give you one more. I will not identify the laboratory or the author, but a certain report was given to me to review critically. It was a detailed analysis of a certain class of properties on a certain material. It was all entered into a computer and had been exquisitely analyzed with all the statistical techniques known to man. Some results came out much like those that Bob Freeman described. On close inspection it turned out that the author had used data as input which were not from the original primary literature, but were numbers out of handbooks and out of other review papers. Thus, they were themselves other averages. With a little further work, it could be deduced that some of the same numbers appeared again and again and again. These had all been considered by this author as independent parts of the same general data population.

TED SELOVER - Bill Thompson and I were talking this noon about the errors that are propagated in textbooks. Some compounds can be found whose heats of formation have the wrong sign. This then gets repeated in a handbook of chemistry and physics, and people start using them for doing heats of reaction which do not make sense. There are many errors in Kubaschewski's book which many consider to be one of the best sources of data. Because the signs are wrong the data are wrong. Maybe an oversight committee is needed to change errors that are found in textbooks.

Well, it looks like we have run out of comments. We want to thank everyone for coming and making this a very successful discussion.

SYMPOSIUM SERIES

ADSORPTION

- 74 Physical Adsorption Processes and Principles
- 96 Developments in Physical Adsorption
- 117 Adsorption Technology
- 134 Gas Purification by Adsorption
- 152 Adsorption and Ion Exchange
- 219 Recent Advances in Adsorption and Ion Exchange
- 230 Adsorption and Ion Exchange—'83
- 233 Adsorption and Ion Exchange—Progress and Future Prospects

AEROSPACE

- 33 Rocket and Missile Technology
- 40 Applications of Plastic Materials in Aerospace
- 52 Chemical Engineering Techniques in Aerospace
- 61 Aerospace Chemical Engineering
- 68 Aerospace Life Support

BIOENGINEERING

- 66 Chemical Engineering in Medicine
- 69 Bioengineering and Food Processing
- 84 The Artificial Kidney
- 86 Bioengineering ... Food
- 93 Engineering of Unconventional Protein Production
- 99 Mass Transfer in Biological Systems
- 108 Food and Bioengineering—Fundamental and Industrial Aspects
- 114 Advances in Bioengineering
- 132 Engineering of Food Preservation and Biochemical Processes
- 158 Biochemical Engineering—Energy, Renewable Resources and New Foods
- 163 Water Removal Processes: Drying and Concentration of Foods and Other Materials
- 172 Food, pharmaceutical and bioengineering—1976/77
- 181 Biochemical Engineering Renewable Sources of Energy and Chemical Feedstocks

CRYSTALLIZATION

- 95 Crystallization from Solution and Melts
- 110 Factors Influencing Size Distribution
- 121 Nucleation Phenomena in Growing Crystals Systems
- 153 Analysis and Design of Crystallization Processes
- 193 Design Control and Analysis of Crystallization Processes
- 215 Nucleation, Growth and Impurity Effects in Crystallization Process Engineering

DRAG REDUCTION

- 111 Drag Reduction
- 130 Drag Reduction in Polymer Solutions

ENERGY

Conversion and Transfer

- 5 Heat Transfer, Atlantic City
- 9 Heat Transfer, Research Papers for 1954
- 17 Heat Transfer, St. Louis
- 18 Heat Transfer, Louisville
- 29 Heat Transfer, Chicago
- 30 Heat Transfer, Storrs
- 32 Heat Transfer, Buffalo
- 41 Heat Transfer, Houston
- 57 Heat Transfer, Boston
- 59 Heat Transfer, Cleveland
- 64 Heat Transfer, Los Angeles
- 75 Energy Conversion Systems
- 79 Heat Transfer with Phase Change
- 82 Heat Transfer—Seattle
- 87 Advances in Cryogenic Heat Transfer
- 92 Heat Transfer—Philadelphia
- 102 Heat Transfer—Minneapolis
- 113 Convective and Interfacial Heat Transfer
- 118 Heat Transfer—Tulsa
- 119 Commercial Power Generation
- 131 Heat Transfer: Fundamentals and Industrial Applications
- 138 Heat Transfer—Research and Design
- 162 Energy and Resource Recovery from Industrial and Municipal Solid Wastes
- 174 Heat Transfer: Research and Application
- 189 Heat Transfer—San Diego 1979
- 199 Heat Transfer—Orlando 1980
- 202 Transport with Chemical Reactions
- 208 Heat Transfer—Milwaukee 1981
- 216 Processing of Energy and Metallic Minerals
- 225 Heat Transfer—Seattle 1983
- 236 Heat Transfer—Niagara Falls 1984

Nuclear Engineering

- 11 Part I (Ann Arbor)
- 12 Part II (Ann Arbor)
- 13 Part III (Ann Arbor)
- 19 Part IV
- 22 Part V
- 23 Part VI
- 27 Part VII
- 28 Part VIII
- 47 Part X
- 51 Part XI
- 53 Part XIII
- 56 Part XIV
- 60 Part XII
- 65 Part XV
- 68 Part XVI
- 71 Part XVII
- 80 Part XVIII
- 83 Part XIX
- 94 Part XX
- 104 Part XXI
- 106 Part XXII
- 119 Commercial Power Generation
- 123 Part XXIII
- 154 Radioactive Wastes from the Nuclear Fuel Cycle
- 164 Nuclear, Solar, and Process Heat Transfer—St. Louis 1976
- 168 Heat Transfer in Thermonuclear Power Systems
- 169 Developments in Uranium Enrichment
- 191 Nuclear Engineering Questions Power Reprocessing, Waste, Decontamination Fusion
- 221 Recent Developments in Uranium Enrichment

ENVIRONMENT

- 35 Pollution and Environmental Health
- 45 Pollution Control Engineering
- 78 Water Reuse
- 90 Water—1968
- 97 Water—1969
- 107 Water—1970
- 115 Important Chemical Reactions in Air Pollution Control
- 122 Chemical Engineering Applications of Solid Waste Treatment
- 124 Water—1971
- 126 Air Pollution and its Control
- 129 Water—1972
- 133 Forest Products and the Environment
- 136 Water—1973
- 137 Recent Advances in Air Pollution Control
- 139 Advances In Processing and Utilization of Forest Products
- 144 Water—1974: I. Industrial Wastewater Treatment
- 145 Water—1974: II. Municipal Wastewater Treatment
- 146 Forest Product Residuals
- 147 Air: I. Pollution Control and Clean Energy
- 148 Air: II. Control of NO_x and SO_x Emissions
- 149 Trace Contaminants in the Environment
- 150 Advances in Interfacial Phenomena of Particulate/Solution/Gas Systems: Applications to Flotation Research
- 151 Water—1975
- 156 Air Pollution Control and Clean Energy
- 157 New Horizons for the Chemical Engineer in Pulp and Paper Technology
- 165 Dispersion and Control of Atmospheric Emissions, New-Energy-Source Pollution Potential
- 166 Water—1976: I. Physical, Chemical Wastewater Treatment
- 167 Water—1976: II. Biological Wastewater Treatment
- 170 Intermaterials Competition in the Management of Shrinking Resources
- 171 What the Filterman Needs to Know About Filtration
- 175 Control and Dispersion of Air Pollutants: Emphasis on NO_X and Particulate Emissions
- 177 Energy and Environmental Concerns in the Forest Products Industry
- 184 Advances in the Utilization and Processing of Forest Products
- 188 Control of Emissions from Stationary Combustion Sources Pollutant Detection and Behavior in the Atmosphere
- 190 Water—1978
- 195 The Role of Chemical Engineering in Utilizing the Nation's Forest Resources
- 196 Implications of the Clean Air Amendments of 1977 and of Energy Considerations for Air Pollution Control
- 197 Water—1979
- 198 Fundamentals and Applications of Solar Energy
- 200 New Process Alternatives in the Forest Products Industries
- 201 Emission Control from Stationary Power Sources: Technical, Economic and Environmental Assessments
- 207 The Use and Processing of Renewable Resources—Chemical Engineering Challenge of the Future
- 209 Water—1980
- 210 Fundamentals and Applications of Solar Energy II
- 211 Research Trends in Air Pollution Control: Scrubbing, Hot Gas Clean-up, Sampling and Analysis
- 213 Three Mile Island Cleanup
- 223 Advances in Production of Forest Products
- 232 Applications of Chemical Engineering in the Forest Products Industry

FLUIDIZATION

- 38 Fluidization
- 62 Fluid Particle Technology
- 67 Fluidized Bed Technology
- 101 Fundamental Processes in Fluidized Beds
- 105 Fluidization Fundamentals and Application
- 116 Fluidization: Fundamental Studies Solid-Fluid Reactions, and Applications
- 128 Fluidized Bed Fundamentals and Applications
- 141 Fluidization and Fluid-Particle Systems
- 161 Fluidization Theories and Applications
- 176 Fluidization Application to Coal Conversion Processes
- 205 Recent Advances in Fluidization and Fluid-Particle Systems
- 234 Fluidization and Fluid Particle Systems: Theories and Applications

HISTORY OF CHEMICAL ENGINEERING
- 100 The History of Penicillin Production
- 235 Diamond Jubilee Historical/Review Volume

ION EXCHANGE
- 14 Ion Exchange
- 24 Adsorption, Dialysis, and Ion Exchange
- 152 Adsorption and Ion Exchange
- 179 Adsorption and Ion Exchange Separations
- 219 Recent Advances in Adsorption and Ion Exchange
- 230 Adsorption and Ion Exchange—'83
- 233 Adsorption and Ion Exchange—Progress and Future Prospects

KINETICS
- 4 Reaction Kinetics and Transfer Process
- 25 Reaction Kinetics and Unit Operations
- 72 Recent Advances in Kinetics
- 73 Kinetics and Catalysis
- 83 Recent Advances in Kinetics and Catalysis

MATHEMATICS
- 31 Advances in Computational and Mathematical Techniques in Chemical Engineering
- 37 Applied Mathematics in Chemical Engineering
- 42 Statistical and Numerical Methods in Chemical Engineering
- 50 Optimization Techniques

MINERALS
- 15 Mineral Engineering Techniques
- 20 Liquid Metals Technology—Part I
- 43 Recent Advances in Ferrous Metallurgy
- 85 Fossil Hydrocarbon and Mineral Processing
- 173 Fundamental Aspects of Hydrometallurgical Processes
- 180 Spinning Wire from Molten Metals
- 216 Processing of Energy and Metallic Minerals

PETROCHEMICALS
- 34 Petrochemicals and Petroleum Refining
- 49 Polymer Processing
- 127 Declining Domestic Reserves—Effect on Petroleum and Petrochemical Industry
- 135 The Petroleum/Petrochemical Industry and the Ecological Challenge
- 142 Optimum Use of World Petroleum
- 212 Interfacial Phenomena in Enhanced Oil Recovery

PETROLEUM PROCESSING
- 34 Petrochemicals and Petroleum Refining
- 54 Hydrocarbons from Oil Shale, Oil Sands, and Coal
- 98 Methanol Technology and Economics
- 103 C_4 Hydrocarbon Production and Distribution
- 127 Declining Domestic Reserves—Effect on Petroleum and Petrochemical Industry
- 135 The Petroleum/Petrochemical Industry and the Ecological Challenge
- 142 Optimum Use of World Petroleum
- 155 Oil Shale and Tar Sands
- 226 Underground Coal Gasification: The State of the Art

PHASE EQUILIBRIA
- 2 Pittsburgh and Houston
- 3 Minneapolis and Columbus
- 6 Collected Research Papers
- 81 Phase Equilibria and Related Properties
- 88 Phase Equilibria and Gas Mixtures Properties

PROCESS DYNAMICS
- 36 Process Dynamics and Control
- 46 Process Systems Engineering
- 55 Process Control and Applied Mathematics
- 159 Chemical Process Control
- 214 Selected Topics on Computer-Aided Process Design and Analysis

SEPARATION
- 91 Unusual Methods of Separation
- 120 Recent Advances in Separation Techniques
- 192 Recent Advances in Separation Techniques—II

SONICS
- 1 Ultrasonics—Two Symposia
- 109 Sonochemical Engineering

THERMODYNAMICS
- 7 Applied Thermodynamics
- 44 Thermodynamics
- 140 Thermodynamics-Data and Correlations

TRANSPORT PROPERTIES
- 16 Mass Transfer
- 56 Selected Topics in Transport Phenomena
- 77 Fundamental Research on Heat and Mass Transfer

MISCELLANEOUS
- 26 Chemical Engineering Education—Academic and Industrial
- 48 Chemical Engineering Reviews
- 70 Small-Scale Equipment for Chemical Engineering Laboratories
- 76 High Pressure Technology
- 112 Engineering, Chemistry, and Use of Plasma Reactors
- 125 Vacuum Technology at Low Temperatures
- 143 Standardization of Catalyst Test Methods
- 160 Continuous Polymerization Reactors
- 182 Biorheology
- 183 The Modern Undergradate Laboratory Innovative Techniques
- 185 Electro Organic Synthesis Technology
- 186 Plasma Chemical Processing
- 187 Chronic Replacement of Kidney Function
- 194 Hazardous Chemical—Spills and Waterborne Transportation
- 203 A Review of AIChE's Design Institute for Physical Property Data (DIPPR) and Worldwide Affiliated Activities
- 204 Tutorial Lectures in Electrochemical Engineering and Technology
- 206 Controlled Release Systems
- 217 New Composite Materials and Technology
- 220 Uncertainty Analysis for Engineers
- 228 Problem Solving
- 229 Tutorial Lectures in Electrochemical Engineering and Technology—II
- 231 Data Base Implementation and Application
- 237 Awareness of Information Sources

MONOGRAPH SERIES
- 1 Reaction Kinetics by Olaf Hougen
- 2 Atomization and Spray Drying by W.R. Marshall, Jr
- 3 The Manufacture of Nitric Acid by the Oxidation of Ammonia—The DuPont Pressure Process by Thomas H. Chilton
- 4 Experiences and Experiments with Process Dynamics by Joel O. Hougen
- 5 Present Past, and Future Property Estimation Techniques by Robert C. Reid
- 6 Catalysts and Reactors by James Wei
- 7 The 'Calculated' Loss-of-Coolant Accident by L.J. Ybarrondo, C.W. Solbrig, H.S. Isbin
- 8 Understanding and Conceiving Chemical Process by C. Judson King
- 9 Ecosystem Technology: Theory and Practice by Aaron J. Teller
- 10 Fundamentals of Fire and Explosion by Daniel R. Stull
- 11 Lumps, Models and Kinetics in Practice by Vern W. Weekman, Jr.
- 12 Lectures in Atmospheric Chemistry by John H. Seinfeld
- 13 Advanced Process Engineering by James R. Fair
- 14 Synfuels from Coal by Bernard S. Lee